INSTRUC

An introduction to
GLOBAL ENVIRONMENTAL ISSUES

Second Edition

Kevin T. Pickering
and
Lewis A. Owen

LONDON AND NEW YORK

First published 1997
by Routledge
11 New Fetter Lane, London EC4P 4EE

Simultaneously published in the USA and Canada
by Routledge
29 West 35th Street, New York, NY 10001

Typeset in Galliard by Florencetype Ltd, Stoodleigh, Devon

Printed and bound in Great Britain by Butler & Tanner Ltd,
Frome and London

British Library Cataloguing in Publication Data
A catalogue record for this book is available from the British Library

Library of Congress Cataloguing in Publication Data
A catalogue record for this book has been requested

ISBN 0-415-16664-0

Contents

The fact of the submersion to that extent being admitted [marine shells up to 360 feet in elevation being discovered in gravels, sands and muds throughout Scotland] *several interesting questions arise. Has the change of the relative level of land and sea been accomplished by an upward movement of the land, or by recession of the sea? Has the shift been slow and equable with regard to time, or by fits and starts with long pauses between, or by a slow movement interrupted by pauses? Has the time embraced by the whole series of phenomena been long or short, geologically speaking? What have been the general and particular circumstances and results of the whole movement?*

Ancient sea-margins as memorials of changes in the relative level of sea and land, Robert Chambers (1848)

Introduction

This manual is designed to aid instructors who have adopted the textbook *An Introduction to Global Environmental Issues* as a course text. It will help in the preparation of lectures, seminars, tutorials and assignments. The manual deals with each of the ten chapters in the core text, presenting the main aims, key points and topics for discussion. It is necessarily concise in order to permit the teacher to assess the contents of each textbook chapter rapidly. Potential problem areas for students are highlighted, and additional material is provided where appropriate.

The manual considers each textbook chapter in the following sections:

Aims

This section provides a list of the main aims of each chapter, thereby enabling the teacher/lecturer/instructor to introduce the various components of a course or topics, and clearly emphasise the rationale behind each chapter to the students.

Key point summary

This is a summary of the main points in each chapter. They allow the instructor to prepare structured lectures and to assess the important issues in each chapter.

Main learning hurdles

This section highlights the major problems that are likely to affect the students' understanding of each chapter. It is divided into several key points to allow the instructor to anticipate and reduce the hurdles that the student may encounter.

Key terms

The key terms within each chapter are listed so that instructors can assess the terminology and familiarise themselves with them before teaching the students.

Issues for group discussion

The issues for group discussion help instructors to prepare structured tutorials and seminars. They aim to stimulate active discussion among staff and students. The instructor should advise the students to read the appropriate chapter and the relevant selected reading(s) before the discussions.

Selected readings

This provides a list of selected papers and articles that appear within each chapter. The readings are from a range of different types of sources, are enjoyable and easy to read, and are present in most libraries.

List of main textbooks

The lists of main textbooks help the instructor to acquire additional information that will complement each chapter or will help to widen the scope of *An Introduction to Global Environmental Issues*.

Essay questions

A list of essay questions will allow the instructor to provide assessments that will encourage the student to read each chapter carefully and access referenced sources. The answers to each of these questions must be less than 1,500 words. The instructor must remind students that the essays must have full references, in the Harvard style, and that figures and tables improve the essay.

Multiple-choice questions

The multiple-choice questions, which vary in their number as seems appropriate, provide a means of informally testing the students' knowledge after they have read each chapter. Instructors must not, however, use these for student assessment, because multiple-choice questions tend to encourage students to concentrate on recall. Furthermore, most questions posed in environmental science are not simple; rather they require considerable discussion. It is, therefore, better if the instructor assesses the students in other ways, as outlined in Chapter 11.

Figure questions

The figure questions relate to the figures in the second edition of *An Introduction to Global Environmental Issues*. The questions are designed to test a student's understanding of some of the critical figures. They also help the students to study the figures in depth as well as highlighting how they should be using the figures in the main text. In addition, several questions accompany each figure, to help test the student's broader knowledge in relation to the figure in question.

Short questions

Ten short questions provide a test of the student's detailed knowledge of each chapter. Each answer should be no more than a paragraph long. Model answers accompany each question at the end of the section.

Additional references

As with any book dealing with global environmental issues, there is a time lag between completion and publication during which many events take place that might be included in the book. This teacher's manual was completed a few months after the second edition of *An Introduction to Global Environmental Issues*, during which time there have been important additional new publications, and older ones have become apparent as worthy of inclusion. This section at the end of each chapter lists references that a teacher may wish to follow up.

Most of the references are unapologetically from the mainstream international science journal, *Nature*, as this is a very accessible reference source for most students taking any course in environmental sciences.

We hope that the manual will stimulate both instructors and students and that the suggestions and information will enthuse.

Aims

- To introduce students to the main global issues.

- To outline the main disciplines involved in environmental science.

- To introduce:
  - The Earth's context in space
  - Life and biodiversity
  - The evolution of life and of the atmosphere
  - Plate tectonic theory
  - Climatology and meteorology
  - The hydrological cycle.

- To examine the concepts of the 'systems approach', thresholds, chaos and rates of change.

Key point summary

- The Earth is merely one of an extremely large number of planets in the known Universe but may be unique in supporting life as we know it. The discovery of the remnants of possible early life-forms in a Martian meteorite analysed in 1996, however, raises the question that life in some form, even if primitive, may be more common than has been widely thought up until now. Most recently (October 1996), scientists have re-examined the remnants of this possible Martian life and concluded that the structures are inorganic.

- The outer layers of the Earth comprise the **atmosphere**, **biosphere**, **hydrosphere** and **lithosphere**. These are interrelated as the **ecosphere**.

- The Earth's interior comprises the **core**, **mantle** and **crust**.

- The upper part of the mantle and crust, the lithosphere, is the most significant with regard to environmental studies. The lithosphere is essentially rigid and rests on the **asthenosphere**, which has the capacity to flow. The movement of lithospheric plates, their construction and destruction, constitutes the study of **plate tectonics**. Plate tectonic theory explains the distribution of Earth's surface features, rocks and resources, and earthquakes and volcanoes.

- Biological diversity is enormous and is sustained by energy from the Sun and the Earth's internal energy systems. **Biodiversity** is essential to maintaining the stability of ecosystems and **biogeochemical cycles**. It is protected by the UN Convention on Biodiversity and the World Conservation Strategy.

- Animals and plants inhabit particular **ecological niches**, **biomes** and **biotic provinces**, which are dominated by a variety of environmental factors directly or indirectly influenced by local, regional and/or global climates.

- Biological activity plays a critical role in releasing free O_2 to the atmosphere by **oxygenic photosynthesis**.

- Evidence from chemical isotopes in sedimentary rocks from Isua in west Greenland suggests that life existed on Earth from at least approximately 3.8 Ga. A study of carbon and **strontium isotopes** in sedimentary rocks suggests that free O_2 started to accumulate in substantial quantities in the Earth's atmosphere about 2,000 million years ago as anoxic basins began to form by plate tectonic processes, which allowed organic carbon to be buried. Prior to this, oxygen was held in carbonate rocks as the so-called **carbonate reservoir**. Oxygenic photosynthesis took place at least 600 million years ago and provided a mechanism capable of sustaining atmospheric free O_2 levels.

- The Sun provides the energy to drive photosynthesis, and atmospheric and hydrological systems.

- On the basis of temperature, the atmosphere divides into a series of layers, the **troposphere**, **stratosphere**, **mesosphere** and **thermosphere**.

Most weather processes occur in the lowest layer, the troposphere.

- The world divides into climatic regions. These are influenced and partially defined by the general atmospheric circulation. This is driven by differential heating and cooling of the Earth's land surface, its oceans and atmosphere, and the Earth's rotation.

- The main components of global atmospheric circulation include the **polar high-pressure cell**, the **circumpolar vortex**, the **Hadley cells** and the **intertropical convergence zone (ITCZ)**.

- The **hydrological cycle** involves the storage and transfer of water throughout the world and involves hydrological, atmospheric, biological and geological processes.

- The 'systems approach' allows the independent study of the various components of the ecosphere, from which it is possible to appreciate both **negative** and **positive** feedbacks.

- The **Gaia Hypothesis** describes the Earth as a self-regulating organism, able to sustain itself in equilibrium with any long-term major changes, maintaining climate, and the composition of the atmosphere, soils and oceans.

- Processes and events within environmental systems may change from one level or condition to another when an input has reached a **critical threshold**.

- **Chaos theory** proposes that natural systems are fundamentally unpredictable.

- Time scales and rates of change vary from hundreds of million of years to fractions of a second. Such rates and magnitudes must be appreciated when considering environmental systems.

- The evolution of life is a continuous process punctuated by periods of rapid change marked by the appearance of new species and the disappearance or extinction of other species. It is these relatively rapid changes in life on Earth that allow a sensible division of geological time.

- Environmental awareness began with the interest in natural history during the nineteenth century. It evolved into the study of ecology and initiated the development of international scientific programmes and organisations in the 1950s. Recently it has concentrated on anthropogenically induced global climate change.

Main learning hurdles

The principles of plate tectonic theory

Plate tectonic theory is one of the more exciting theories to explain to students. Some basic physical principles must be explained for the students to maximise their understanding of the theory. These include the internal structure of the Earth, heat convection, the Earth's magnetism, the properties and characteristics of solid and molten rock, and the rates of tectonic processes, together with an appreciation of the geological time scale.

The dynamics of the climatic system

The instructor must help the students to examine the various components of the climatic system. This involves an appreciation of some basic physical principles, such as radiation, heat exchange, and temperature and humidity. Care must be taken to describe these before outlining the dynamics of heat exchange and global atmospheric circulations.

Gaia

The instructor must explain the evolution of Gaia in terms of scientific theory so that students can assess the validity of Gaia. The biogeochemical cycles must be explained so that the students can fully understand Gaia and use it as a framework for examining environmental change.

The geological time scale

This can be a difficult concept for students to understand. The instructor must emphasise the range of scales over which processes operate and put these into context with how Earth scientists are involved in examining the rates and magnitudes of change, and why geological time is divided in the way that it is. The instructor must explain the various divisions of the geological time scale, for example periods, eras, epochs and stages.

Key terms

Anthropogenic; asthenosphere; Big Bang; bioherm; biosphere; biota; biotope; butterfly effect; core;

Coriolis effect; cybernetics; ecosystem; electromagnetic spectrum; eukaryotic organism; feedback; fossil fuel; fractal geometry; geodesy; geosphere; greenhouse gas; greenhouse effect; Hadley cell; hominid; hot spot; hydrology; hydrosphere; intertropical convergence zone; lipid; lithosphere; mantle; meteorology; mid-ocean ridge; orogeny; oxygenic photosynthesis; pedology; Proterozoic; punctuated evolution; Rossby wave; sea-floor spreading; stratosphere; tectonic processes; troposphere; westerlies.

Issues for group discussion

Discuss the relevance of Gaia to environmental science. Following the formulation and illustration by examples of Lovelock's ideas of Gaia, instructors and students should discuss Lovelock's ideas in terms of whether they constitute a testable theory, a hypothesis or merely a paradigm. The discussion should also examine various alternative arguments, such as those of Des Morais *et al.* (1992). Des Morais *et al.* (1992), for example, propose that geological processes are more important in controlling the evolution of the atmosphere. The discussion should develop the interlinks between the biotic and abiotic world.

Discuss the meaning of rapid change.
The students should consider the various types of change, focusing on geological, biological and climatological changes. The discussion should use the geological time scale to discuss these changes in terms of what we are able to experience in our lifetimes.

Examine the controls on the distribution of animals and plants throughout the world.
The teacher should encourage the students to compare tectonic, climatological, oceanic and vegetation maps of the world. They should continue by discussing the controls on climate, and the evolution of the continents that have led to distinct biomes. The discussion should also consider the role of human extermination of organisms. This discussion should help students to think geographically and to consider the interaction between various systems that make up the ecosphere.

Selected readings

England, P. 1992. Deformation of the continental crust. In: Brown, G.C., Hawskesworth, C.J. and Wilson, R.C.L. (eds), *Understanding the Earth: A New Synthesis*, 275–300. Cambridge: Cambridge University Press.
This chapter provides a good summary of the main current views on crustal deformation. It is easy to follow, given a basic knowledge of plate tectonics.

Myers, N. 1990. The biodiversity challenge: expanded hotspot analysis. *The Environmentalist*, 10 (4), 1–14.
This article provides a good analysis of the importance of biodiversity and the concept of hot spots. Very easy to read and well referenced.

Myers, N. 1995. Environmental unknowns. *Science*, 269, 358–60.
This is an interesting paper that discusses the importance of directing research towards areas that might produce new threats to the environment, e.g. environmental surprises. It discusses two sets of environmental surprises: environmental discontinuities, the result of ecological systems jumping over a threshold; and synergisms, the result of two or more environmental processes interacting in such a way that the outcome is not additive but multiplicative.

Pimm, S.L., Russell, G.J., Gittleman, J.L. and Brooks, T.M. 1995. The future of biodiversity. *Science*, 269, 347–50.
This is an interesting paper discussing the importance of biodiversity and conservation strategies.

Textbooks

Botkin, D. and Keller, E. 1995. *Environmental Science: Earth as a Living Planet*. Chichester: John Wiley & Sons Ltd.
This is a colourful and well-illustrated introductory book for high school and university students on the principles of environmental science. The text is simple to follow, aided by case studies and explanations in boxed text, and a series of appendixes. It contains eight sections: environment as an idea; Earth as a system; life and the environment; sustainable living resources; energy; water environment; air pollution; and environment and society.

Bradshaw, M. and Weaver, R. 1993. *Physical Geography: An Introduction to Earth Environments*. London: Mosby.
This is a comprehensive and well-illustrated textbook outlining the principles of Earth systems at an introductory level suitable for high school students and first-year undergraduates. The book describes atmosphere–ocean systems in terms of their dynamics; plate

tectonics; geomorphological processes; aspects of human interaction with the natural environment, and ecological systems, in which there is a useful emphasis on soil dynamics and the characteristics of biomes.

Broecker, W.S. 1987. *How to Build a Habitable Planet.* Palisades, New York: Eldigio Press.
This is an extremely readable introduction to the origin and evolution of the Earth. Broecker manages to make seemingly complex scientific arguments simple and interesting. This book is highly recommended as an introductory book for both students and instructors wishing to understand some basic geochemical arguments about the Earth. Obtaining copies can be a frustrating process as the publishing house does not have a sophisticated global distribution system; therefore, it is recommended that instructors/lecturers order sufficient quantities well before the start of any appropriate course.

Brown, G.C., Hawskesworth, C.J. and Wilson, R.C.L. 1992 (eds). *Understanding the Earth: A New Synthesis.* Cambridge: Cambridge University Press.
This is a useful textbook written for the British Open University. It is a compilation of chapters by some of the leading Earth scientists summarising contemporary studies and issues in geology. It is easy to read and helped by good figures and boxed text. It is essential reading for anyone studying geology.

Dawkins, R. 1986. *The Blind Watchmaker.* Harlow: Longman.
An examination of the evolution of life, which inspires the reader with a vision of existence and the elegance of biological design and complexity. Dawkins argues for the truism of Darwinian theory and shows for example how modern views such as punctuated evolution are part of neo-Darwinian theory. A useful supplementary book for many courses in the natural sciences and environmental studies.

Gleick, J. 1987. *Chaos.* Bungay, Suffolk: Richard Clay Ltd.
This is a readable account of the historical development and the elementary principles of the science of chaos.

Goudie, A. 1993. *The Nature of the Environment,* third edition, Oxford: Blackwell.
This is a comprehensive introduction to the world's natural environments. It examines the dynamics of the processes acting on the landscape and environment, past, present and future, integrating the study of landforms, climate, soils, hydrology, plants and animals to provide a greater understanding of the nature of environments both on the global and local scales.

Huggett, R.J. 1995. *Geoecology: An Evolutionary Approach.* London: Routledge.
A refreshing text examining the dynamics of geoecosystems. It develops a model for geoecosystems, organised on a hierarchical basis, which responds continuously to changes within themselves and the near-surface environment (atmosphere, hydrosphere and lithosphere), as well as geological and cosmic influences.

Jackson, A.R.W. and Jackson, J.M. 1996. *Environmental Science: The Natural Environment and Human Impact.* Harlow: Longman.
This is a good undergraduate introductory text to environmental science exploring the fundamental concepts of the natural environment, and the interactions between the lithosphere, hydrosphere, atmosphere and biosphere. The book emphasises the consequences for the environment of natural resource exploitation.

Lovelock, J.E. 1988. *The Ages of Gaia: A Biography of our Living Earth.* Oxford: Oxford University Press.
The follow-up book to *Gaia: A New Look at Life on Earth* (1982), which elaborates on the Gaia view of Earth. This book examines the interaction between the atmosphere, oceans, the Earth's crust, and the organisms that evolve and live on Earth. Lovelock discusses recent scientific developments, including those on global warming, ozone depletion, acid rain and nuclear power. This book provides a thought-provoking look at interdependence, and the role of negative and positive feedbacks in controlling the evolution and adaptability of life.

Manahan, S.E. 1993. *Fundamentals of Environmental Chemistry.* Michigan: Lewis Publishers.
A comprehensive and well-written textbook aimed at students having little or no background in chemistry. This book gives the fundamentals of chemistry and environmental chemistry needed for a trade, profession or curriculum of study requiring a basic knowledge of these topics. It also serves as a general reference source. This book will appeal to those involved in college and university studies where the environmental course has a relatively strong science base but is unlikely to appeal to those in the social sciences and geography.

Nebel, B.J. and Wright, R.T. 1993. *Environmental Science: The Way the World Works*, fourth edition. Englewood Cliffs, New Jersey: Prentice Hall.
This is a useful textbook where the central theme is sustainable development. There are four sections in this book: Part I, What ecosystems are and how they work; Part II, Finding a balance between population, soil, water and agriculture; Part III, Pollution; Part IV, Resources: biota, refuse, energy and land. The text has various elements that provide teaching aids, e.g. learning objectives, review questions, etc. A detracting feature is the presentation style of very well-drawn and sophisticated diagrams alongside over simplistic, naïve, artwork. The book aims at college students taking environmental courses.

Summerfield, M.A. 1991. *Global Geomorphology*. Harlow: Longman, 537 pp.
A useful, comprehensive textbook on geomorphology ideal for everyone interested in the Earth's surface and internal processes. The book is beautifully illustrated, with useful tables and boxed text.

Watts, S. (ed.) 1996. *Essential Environmental Science*. London: Routledge.
This practical manual explores the vast range of techniques, methods and basic tools necessary for the study of the environment. This is a useful manual for practitioners and students at all levels.

Yearley, S. 1992. *The Green Case: A Sociology of Environmental Issues, Arguments and Politics*. London: Routledge.
This is a comprehensive account of the basis of 'green' arguments and of their social and political implications. Yearley examines the reasons for the success of leading campaign groups (such as Greenpeace), and analyses developments in green politics and green consumerism. He also emphasises the serious ecological problems in the developing world, and argues that these problems are inextricably linked with debt and their need for development. This is a well-written sociological perspective, and a recommended supplementary book for those interested in the broader aspects of global environmental issues.

Essay questions

1 Assess the assumption that Gaia provides a useful framework for examining contemporary environmental problems.

2 Explain how the lithosphere, biosphere, hydrosphere and atmosphere collectively contribute towards a habitable planet.

3 Describe the controls on the global distribution of biota.

4 Assess the various opinions regarding the role of organisms in controlling the evolution and maintenance of the Earth's atmosphere–ocean system as habitable for life.

5 How do the Earth's lithosphere, hydrosphere and atmosphere processes sustain biological systems?

6 In your opinion, what is the most important global environmental issue that should be addressed by the international scientific community? Justify your choice.

7 The 'systems approach' is a convenient way of studying natural processes, and the nature and rates of change. Describe the systems approach, and illustrate your answer by describing specific natural systems.

8 'Chaos and unpredictability are inherent attributes of all natural systems. Therefore there is no point in trying to understand the complexity of ecosystems with a view to predicting the potential impacts that human activities may have on them.' Discuss.

9 Describe the possible regional and global consequences of a reduction in biodiversity.

10 Describe the main components of general circulation in the atmosphere and its role in defining the world's climatic and vegetation regions.

11 Evaluate the factors that have affected the distribution of plants and animal species in the temperate mid-latitudes of the Northern Hemisphere.

12 Assess the relative importance of the main types of events, quasi-periodic and periodic processes that help to drive global climate change over different durations of geological time.

Multiple-choice questions

Choose the best answer for each of the following questions.

1 The age of the Earth is approximately:
(a) 4,600 thousand years
(b) 4,600 million years
(c) 4,600 billion years
(d) 4,600 years

2 Biodiversity is essential:
(a) to sustain the stability of ecosystems
(b) as a gene pool
(c) as an important component of biogeo-chemical cycles
(d) a, b and c

3 The proportion of the Earth's atmosphere in the lowest 5 km is:
(a) 10 per cent
(b) 50 per cent
(c) 75 per cent
(d) 90 per cent

4 The Mohorovicic discontinuity (Moho) separates the Earth's:
(a) mantle from the crust
(b) core from the mantle
(c) asthenosphere and mantle
(d) lithosphere from the crust

5 The asthenosphere is
(a) a zone of partial melting in the upper mantle
(b) a semi-rigid zone of the upper mantle
(c) a fluid zone in the lower crust
(d) a fluid zone in mountain ranges

6 The theory of continental drift was proposed by:
(a) Charles Darwin
(b) Alfred Wallace
(c) Alfred Wegener
(d) Sherlock Holmes

7 Across mid-ocean ridges, palaeomagnetic anomalies form patterns that are:
(a) symmetric
(b) asymmetric
(c) irregular
(d) concentric

8 The Wadati–Benioff zone is a zone of seismicity associated with:
(a) subduction zones
(b) oceanic spreading ridges
(c) convergent plate boundaries
(d) transform boundaries

9 The global ecosystem can be divided into characteristic regional ecosystems in which creatures have evolved to adapt to the climate and topography. These are known as:
(a) ecospheres
(b) niches
(c) biomes
(d) biosphere

10 The processes in which biological activities result in the release of O_2 to the atmosphere are known as:
(a) oxidation
(b) oxygenic photosynthesis
(c) respiration
(d) biosynthesis

11 Hadley cells are:
(a) plant cells that photosynthesise
(b) large-scale atmospheric convective systems
(c) flows of molten rock within the lithosphere
(d) pockets of tropical rainforest with exceptional biodiversity

12 The proportion of carbon dioxide in the Earth's atmosphere is approximately:
(a) 20 per cent
(b) 80 per cent
(c) 3 per cent
(d) 0.3 per cent

13 Positive feedback systems result in:
(a) an increase in the output of the system
(b) a decrease in the output of the system
(c) an increase or decrease in the output of the system
(d) a, b and c

14 The Gaia Hypothesis proposes that:
(a) geological activity is responsible for regulating the composition of the atmosphere
(b) the Earth is a self-regulating system
(c) the Earth's climate is regulated by changes in the Sun's temperature
(d) a, b and c

15 In biological terms, a 'hot spot' is an area that has:
(a) extreme weather conditions
(b) exceptionally high concentrations of species
(c) rapid population growth
(d) a, b and c

16 Most weather processes operate in the:
(a) stratosphere
(b) mesosphere
(c) lithosphere
(d) troposphere

17 The normal environmental lapse rate in the lower atmosphere is approximately:
(a) $65°C\ km^{-1}$
(b) $6.5°C\ km^{-1}$
(c) $0.65°C\ km^{-1}$
(d) $0.65°C\ m^{-1}$

18 The term used to describe the electromagnetic radiation that the Earth receives from the Sun is:
(a) long-wave radiation
(b) insolation
(c) heat
(d) infrared radiation

19 The property that describes the ability of a material to reflect the solar radiation is known as:
(a) insolation
(b) shine
(c) albedo
(d) absorption

20 Temperature differences create pressure gradients, which produce:
(a) rain
(b) winds
(c) ocean currents
(d) waves

21 The rotation of the Earth leads to the deflection and acceleration of many winds in a clockwise direction in the Northern Hemisphere and anticlockwise in the Southern Hemisphere. This effect is known as the:
(a) Wallace effect
(b) Geostrophic effect
(c) Rotation effect
(d) Coriolis effect

22 The great circumpolar flow of winds around each hemisphere, produced by colder air and high pressures in the polar regions, is known as the:
   (a) temperate jetstream
   (b) trade winds
   (c) circumpolar vortex
   (d) intertropical convergence zone

23 Throughout geological history, major extinction event(s) have occurred at the:
   (a) boundary between the Cretaceous and Tertiary
   (b) boundary between the Permian and Triassic
   (c) during late Devonian time
   (d) a, b and c

24 Strontium isotopes in marine carbonates are used to provide information regarding:
   (a) changes in the bioproductivity of the oceans
   (b) changes in ocean temperature
   (c) changing sediment supplies from various sources
   (d) changing sea levels

25 The first modern humans (*Homo sapiens sapiens*) appeared about:

   (a) 20 million years ago
   (b) 2 million years ago
   (c) 200,000 years ago
   (d) 20,000 years ago

Figure questions

1 Figure 1.4 shows a schematic section across an ocean and a continental margin. Answer the following questions.
   (a) What type of plate margins are present?
   (b) Giving reasons for your answer, justify where you believe the oldest oceanic crust is on the diagram.
   (c) Outline the main processes that drive the motion of tectonic plates.

2 Figure 1.12 shows the principal components of a climate system. Answer the following questions.
   (a) How can changes in the extent of glacial ice and vegetation cover affect the climate system?
   (b) Outline the various ways in which the extent of ocean cover can change with time.
   (c) Describe the role of clouds in the climate system.

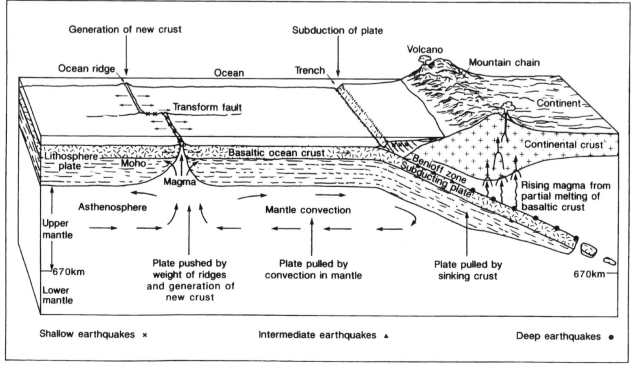

Figure 1.4   *The structure of the outer layers of the Earth, the major physiographic features, and the plate tectonic mechanisms responsible for the generation of new sea-floor crust, sea-floor spreading, the consumption or subduction of oceanic crust, earthquakes and vulcanicity. Redrawn after Selby (1985).*

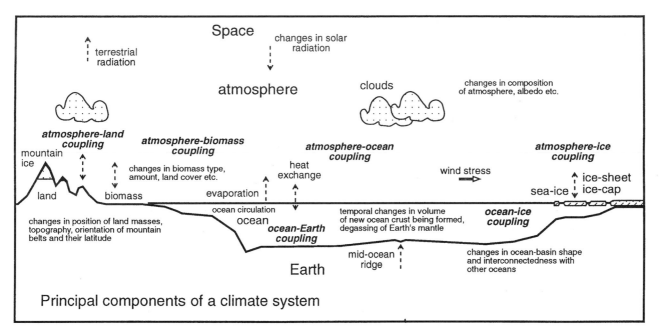

Figure 1.12 *Principal components of a climate system.*

3 Figure 1.15A is a schematic representation of the carbon cycle. Answer the following questions.

(a) Annotate the diagram to show the main ways in which carbon is stored in rocks.

(b) Explain how a rise in sea level may affect the carbon cycle.

(c) What other ways can carbon be added to the atmosphere that are not illustrated on the diagram?

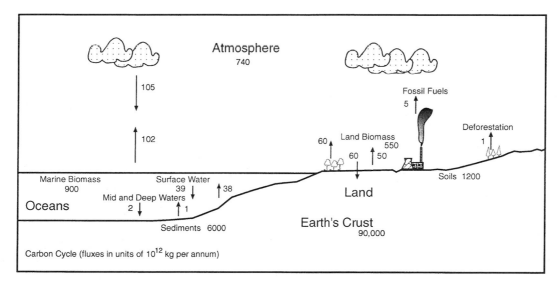

Figure 1.15A *The carbon cycle.*

Short questions

1 Describe the main characteristics of oceanic and continental crust.

2 Describe the nature of the palaeomagnetic anomalies that Vine and Matthews considered in 1963.

3 Why is biodiversity important?

4 What is a biotic province?

5 Describe and provide the appropriate chemical equation for oxygenic photosynthesis.

6 In terms of biodiversity, what is a 'hot spot' and why are they important?

7 What is a jetstream?

8 Outline the basic components and flow of water within the hydrological cycle.

9 Describe the concept of thresholds and why they are important in environmental science.

10 What is the difference between positive and negative feedback?

Answers to multiple-choice questions

1 b; 2 d; 3 b; 4 a; 5 a; 6 c; 7 a; 8 a; 9 c; 10 b; 11 b; 12 a; 13 c; 14 b; 15 b; 16 d; 17 b; 18 b; 19 c; 20 b; 21 d; 22 c; 23 b; 24 c; 25 c.

Answers to figure questions

1 (a) Constructive plate margin (spreading ridge); destructive plate margin (oceanic–continental subduction zone). (b) The oldest oceanic crust is present in the far right of the diagram at a depth of *c.* 670 km. This is the oldest because new ocean crust forms at the spreading ridge, moving away from the ridge with time to be subducted under other plates. (c) The force due to mantle convection and the gravitational drag/pull of the subducting plates drive plate motion 'slab pull' by subduction, and 'slab push' by the creation of new oceanic crust at spreading centres.

2 (a) Vegetation and ice cover affect the albedo, evaporation, ground temperatures, soil moisture and surface properties such as roughness; therefore, changes in these result in changes in heat budget, the hydrological cycle and wind systems. (b) The extent of ocean cover may change due to changes in global temperature; sea level drops as water is stored as ice during cold times and rises during warm times as ice melts and water thermally expands. In addition sea levels rise during times of active sea-floor spreading as ocean ridges rise. (c) Clouds are probably the most important self-regulating mechanism in the ocean–atmosphere system. Their warming effect can be great, but their overall cooling effects make the Earth 10-15 °C cooler than it would otherwise be if it were cloudless. Tropical clouds reflect sunlight back into the atmosphere to cool the system but also have a greenhouse warming effect. The middle- to high-latitude clouds, however, have a net cooling effect. Clouds also provide extra hydroxyl molecules, which are capable of oxidising CH_4 and NO_x, thereby removing some of the greenhouse gases from the atmosphere and reducing their greenhouse effect.

3 (a) See Figure 1.15. (b) A rise in sea level may affect the carbon cycle by reducing the amount of vegetation and hence the biomass may decrease, which would lead to increased levels of atmospheric CO_2. (c) Carbon is also added to the atmosphere by volcanic activity and metamorphism of rocks.

Answers to short questions

1 Oceanic crust comprises dense rock composed of predominantly iron and magnesium silicate minerals and varies in thickness from 5 to 10 km. While continental crust comprises less dense rock composed of predominantly aluminium silicate minerals, it has a maximum thickness of about 70 km, but averages 35 km in thickness.

2 The palaeomagnetic anomalies described by Vine and Matthews in 1963 occurred across mid-ocean ridges. The magnetic field varied symmetrically from areas with more than (positive anomalies) or less than (negative anomalies) the average magnetic field for the region. Vine and Matthews proposed that they formed when magnetic minerals aligned themselves parallel to the Earth's magnetic field as lavas crystallised shortly after being extruded from mid-oceanic ridges. They suggested that the Earth's magnetic field reverses periodically. This produces negative and positive anomalies, and the newly formed ocean crust is pushed away from the ridge each time new lava erupts to form new crust.

3 Biodiversity is important because an abundance of species provides a buffering against environmental change; different species in an ecosystem have important functions in terms of nutrient cycles and food chains; and biodiversity helps to preserve an abundant and varied gene bank, which is an important resource.

4 A biotic province is a region that is distinguished by a characteristic set of taxa, which have a common genetic heritage, and is confined by barriers that inhibit the spread of the distinct taxa into or from other biotic provinces.

5 Oxygenic photosynthesis is the critical role played by organisms in releasing O_2 to the atmosphere. It involves the splitting of a water molecule to release pure oxygen as outlined in the following reaction:

$$CO_2 + H_2O \rightarrow CH_2O + O_2$$

6 Hot spots are areas that have exceptionally high concentrations of species with high levels of endemism and face exceptional threats of destruction. They are particularly important because they constitute a small area of the Earth's land surface ($c.$ 0.5 per cent), yet have some 20 per cent of the Earth's total plant species. These areas may be protected at relatively little effort to save an extremely large number of species from extinction.

7 Jetstreams are broad belts of upper atmosphere flow. They are narrow bands of rapidly moving air and occur in a zone of steep temperature and pressure gradients at the tropopause. Jetstreams travel in excess of 160 km hr^{-1}. The influence of these jetstreams extends into the lower atmosphere, affecting weather conditions.

8 The hydrological cycle involves the movement of water around the Earth. Water precipitates from the atmosphere as rain or snow, falling on the land and oceans. Some of this water percolates into the soil and bedrock as ground water, commonly towards the sea, and some of the water flows via rivers to lakes and oceans. The amount of precipitation counterbalances the evaporation of water from seas and lakes, from the soil by direct evaporation or drawn up by plants and then released during transpiration. Once the water evaporates, it rises into the atmosphere as vapour, until it condenses and returns to the Earth as precipitation.

9 A threshold describes the level at which some process(es) or event(s) will occur abruptly as a consequence of some input process. It is important to determine the threshold for a particular process to take appropriate prevention, preparedness and mitigation measures to reduce environmental damage.

10 A feedback is the result of an output having an effect on the original input within an environmental system. Positive feedback results in magnification or a decrease in the original input, whereas negative feedback results in the stabilisation of the input.

Additional notes on the evolution of the Earth's ocean–atmosphere system

Teachers at a more advanced level might wish to pursue the arguments relating to the development of the Earth's internal structure, and the early Earth's atmosphere and ocean evolution. These notes are intended as a starting point for such studies.

The formation or accretion of the Earth and Moon within the solar nebula is thought to have taken 50–100 million years, with $^{182}Hf/^{182}W$ isotopic ratios from iron meteorites and lunar basalt suggesting that the formation of the core occurred at least 62 ± 10 million years after the iron meteorites formed (Lee and Halliday 1996).

The differentiation of the Earth's interior into a core and mantle is discussed by Li and Agee (1996). Theoretical considerations have led O'Nions (1996) to propose that after the Earth's core formation at about 4.4 Ga, the mantle became convectively layered with material exchange between the lower and upper mantle becoming small at <1 per cent of the lower mantle mass. Between 3 Ga and 2.5 Ga, there appears to have been a rapid growth and stabilisation of the continents (buoyed up by the widespread emplacement of potassic-rich granites), and in the volume of sediments preserved on this relatively young continental crust, together with a large increase in the areal extent of continental shelves and global climate change (Breuer and Spohn 1995).

There is a large degree of uncertainty attached to the time when the mantle became convectively layered into an upper and lower mantle (separated by the 670-km discontinuity). The probable mantle compositional (spinel–perovskite) phase transition at 670 km appears to stabilise into two-layer convection at Raleigh numbers $>10^7$, whereas at low Re $\leq 10^7$ convection currents easily penetrate the boundary such that the convective cells involve the entire mantle (Breuer and Spohn 1995 and references therein). It has been proposed that the change, at about 2.5 Ga, from the Proterozoic to Archaean reflected the transition from a whole-mantle convection to a convectively layered mantle, with the corollary of plate tectonics being initiated at about 2.5 Ga, i.e. from Archaean time onwards (*ibid.*).

In a study of Fe minerals from subaerially weathered Archaean basalt rinds – ancient soil profiles or paleosols – Rye *et al.* (1995) have constrained atmospheric partial pressures for carbon dioxide, $p\mathrm{CO_2}$, to $< 10^{-1.4}$ atm., equivalent to about 100 times present-day levels of 360 ppmbv, and at least five times lower than the concentrations required by one-dimensional climate models to compensate for the lower solar luminosity at 2.75 Ga. Rye *et al.* (1995) and Holland (1996) therefore propose that if the early Earth's atmosphere contained sufficient $\mathrm{CH_4}$ then this greenhouse gas might have offset the low atmospheric $\mathrm{CO_2}$ concentrations to provide an equable surface temperature, particularly in a world where solar luminosity was less than at present. The presence of atmospheric $\mathrm{CH_4}$ is consistent with the presence of isotopically light carbon in pre-2.25 Ga organic matter (*ibid.*).

Atmospheric $\mathrm{O_2}$ levels appear to have increased abruptly at *c.* 2.05–2.25 Ga (Holland 1996), possibly linked to the progressive decrease in the exhalative volcanic processes causing the injection of gases into the atmosphere, the evolution of cyanobacteria capable of generating free $\mathrm{O_2}$ in an $\mathrm{H_2S}$-rich environment and/or the evolution of cyanobacteria that produced free $\mathrm{O_2}$ in fresh-water lakes. In marine environments, $\mathrm{O_2}$ generation by cyanobacteria may be suppressed by the $\mathrm{H_2S}$ released during the bacterial decomposition of marine organic matter. About 1.05–0.64 Ga, the evolution of non-photosynthetic sulphide-oxidising bacteria appears to have paralleled the large shift in isotopic composition of biogenic sedimentary sulphides (Canfield and Teske 1996). It has been proposed that both events were probably driven by the rise in atmospheric $\mathrm{O_2}$ levels to more than 5–18 per cent of present values (*ibid.*).

Students should be encouraged to find out about the evolution of the Earth's atmosphere and oceans. They should be taught that the concentrations of atmospheric gases, and elemental and ionic abundances in the oceans, evolved to present levels. Students should be encouraged to discuss the arguments for the evolution of an oxygenated atmosphere and how this might have affected life on Earth. Finally, students should examine the arguments about any cyclicity in global climate and biological evolution–extinction, perhaps centred around the so-called Fischer supercycles (e.g. Fischer 1982, 1984), as shown in Figure 1.8.

Breuer, D. and Spohn, T. 1995. Possible flush instability in mantle convection at the Archaean–Proterozoic transition. *Nature*, 378, 608–10.

Canfield, D.E. and Teske, A. 1996. Late Proterozoic rise in atmospheric oxygen concentrations inferred from phylogenetic and sulphur-isotope studies. *Nature*, 382, 127–32.

Fischer, A.G. 1982. Long-term climatic oscillations recorded in stratigraphy. In: *Climate in Earth History*, 97–104. Washington, DC: National Academy Press.

Fischer, A.G. 1984. The two Phanerozoic supercycles. In: Berggren, W.A. and van Couvering, J.A. (eds), *Catastrophes and Earth History*, 129–50. Princeton, NJ: Princeton University Press.

Holland, H.D. 1996. Toward a theory of atmospheric evolution. In: *The History of Degassing of the Earth*. Colston Research Symposium, 29–31 August 1996. Abstract volume.

Lee, D.-C. and Halliday, A.N. 1996. Hafnium–tungsten chronometry and the timing of terrestrial core formation. *Nature*, 378, 771–4.

Li, J. and Agee, C.B. 1996. Geochemistry of mantle–core differentiation at high pressure. *Nature*, 381, 686–9.

O'Nions, R.K. 1996. Earth degassing: 4,500 million years to present. In: *The History of Degassing of the Earth*. Colston Research Symposium, 29–31 August 1996. Abstract volume.

Rye, R. 1996. An upper limit on early atmospheric carbon dioxide levels. In: *The History of Degassing of the Earth*. Colston Research Symposium, 29–31 August 1996. Abstract volume.

Rye, R., Kuo, P.H. and Holland, H.D. 1995. Atmospheric carbon dioxide concentrations before 2.2 billion years ago. *Nature*, 378, 603–5.

Additional references

Blum, J.D. and Erel, Y. 1995. A silicate weathering mechanism linking increases in marine $^{87}\mathrm{Sr}/^{86}\mathrm{Sr}$ with global glaciation. *Nature*, 373, 415–8.
This study of Sr chemical weathering from granitoid soils on alpine glacial moraines in the Wind River Range, Wyoming, shows that during glaciations the global fluviatile $^{87}\mathrm{Sr}/^{86}\mathrm{Sr}$ can increase by an average 0.0002.

Croswell, K. 1996. The Milky Way. *New Scientist*, 25 May, Inside Science, 90, 4 pp.

Erwin, D.H. 1996. The mother of mass extinctions. *Scientific American*, July issue, 56–62.

Fukugita, M., Hogan, C.J. and Peebles, P.J.E. 1996. The history of the galaxies. *Nature*, 381, 489–95.

Gordon, A.L. and Fine, R.A. 1996. Pathways of water between the Pacific and Indian oceans in the Indonesian seas. *Nature*, 379, 146–9.

Gribbin, J. 1995. Structure of the Earth's atmosphere. *New Scientist*, 9 December, Inside Science 86, 4 pp.

Watts, S. and Halliwell, L. (eds) 1996. *Essential Environmental Science: Methods and Techniques*. London: Routledge.

Aims

- To examine the magnitudes and rates of natural climate change.

- To assess the possible driving mechanisms for climate change.

- To examine the detailed climate change during Late Quaternary time.

- To explore the nature of catastrophic climate change.

Key point summary

- The Earth's climate has changed throughout geological time, with six major cold periods called **Ice Ages**.

- Since 2.5 million years ago, global climate has cooled considerably, and the Earth entered its present Ice Age, referred to as the **Quaternary Period**.

- Natural causes of global climate change are the result of:
  (1) tectonic processes – redistributing landmasses and altitude, and volcanic activity changing atmospheric aerosols and gases.
  (2) external processes – sunspot activity and variations in the Earth's orbital parameters (**Milankovitch cyclicity**).
  (3) catastrophic events – meteorite impacts.

- **Palaeoclimatology** is the study of past climates. It uses many types of methods and techniques to reconstruct past climate. These include:
  (1) **petrologic techniques** – the interpretation of sediment and rock.
  (2) **chemical methods** – stable isotope studies, particularly oxygen, carbon and nitrogen isotopes; trace metals.
  (3) **dendrochronology** – the study of tree rings.
  (4) **palaeontology** – particularly the study of pollen, beetles and diatoms.
  (5) **geomorphology** – the study of landforms.

- The Quaternary Period is the most important geological period to study past climate change because it provides the best-preserved evidence for climate change and the Earth is still experiencing Ice Age conditions.

- Climate during Quaternary times has fluctuated from cold (**glacial**) stages, lasting 100,000 to 200,000 years, to warm (**interglacial**) stages, lasting 10,000 to 20,000 years. Less warm (**interstadial**) and cold (**stadial**) periods lasted for a few hundred to thousands of years. Milankovitch cyclicity mainly controls these changes.

- The last glacial–interglacial cycle is the best-studied time period. Isotopic studies, particularly the $\delta^{18}O$ record from the GRIP Summit Ice Core, as well as other types of proxy data, show that climatic changes during this period were extremely rapid.

- A major extinction event, which included the dinosaurs, took place about 65 million years ago at the **Cretaceous–Tertiary (K–T) boundary**. This major event is probably the result of one or more meteorites colliding with the Earth. The impact of the meteorite(s) caused global fires, enhanced levels of atmospheric aerosols and reduced sunlight, which in turn led to global cooling and emission of poisonous chemicals.

- Other mass extinctions have occurred throughout geological time. The greatest occurred about 250 Ma, at the close of the Permian Period and start of the Triassic Period, which involved 95 per cent of all living species. This was not associated with a meteorite impact but rather the growth of a supercontinent in low/equatorial latitudes, which caused a dramatic reduction in the area of favourable ecological niches.

- Plate-tectonic processes can cause spectacular regional changes in climate. About 5–6 Ma, the Mediterranean Sea became landlocked as a result of plate tectonic processes. The Atlantic Ocean waters were sealed off from the Mediterranean Sea. The Mediterranean Sea began to evaporate and dry up to form a desert. This resulted in the accumulation of salts and evaporite minerals, which are approximately 1 km thick. The Mediterranean Sea was periodically replenished by catastrophic flooding from the Atlantic Ocean to about 5 Ma, when it flooded back to conditions similar to today.

Main learning hurdles

Milankovitch cycles

It is commonly difficult for students to envisage how the Earth revolves around the Sun, and the Earth's obliquity and eccentricity of orbit. It is best, when possible, to use props such as a globe and a table lamp to illustrate the orbital changes. Students also find it difficult to envisage how orbital variations can be summed to result in Milankovitch cycles. The instructor should use Figure 2.12 to help to illustrate the effect.

Thermohaline conveyor

Students may find it difficult to understand the dynamics of the thermohaline conveyor because of their lack of understanding of a few basic physical principles. The instructor should, therefore, explain how temperature and salinity of water results in density changes, which control the rise or descent of different bodies of water within the oceans. Furthermore, the instructor must illustrate how these density changes produce the main bottom currents and contribute to the thermohaline conveyor.

Oxygen and other stable isotopes as proxies for climate change

Many students find it difficult to grasp the underlying chemical principles of oxygen isotopes. This is generally because they do not understand the nature of isotopes. The instructor should, therefore, explain isotopes and emphasise that lighter isotopes of oxygen are evaporated more readily and are transported further. Once this has been explained it is easier for students to work from first principles with isotopes associated with sea-level changes and the growth of ice sheets, and hence how they relate to climate.

The internationally accepted stable isotope standards for hydrogen, carbon, nitrogen, oxygen and sulphur are as follows:

Hydrogen	Standard Mean Ocean Water	SMOW
Carbon	*Belemnitella americanus* from the Cretaceous Peedee Formation, South Carolina	PDB
Nitrogen	Atmospheric N_2	–
Oxygen	Standard Mean Ocean Water	SMOW
	Belemnitella americanus from the Cretaceous Peedee Formation, South Carolina	PDB
Sulphur	Troilite (FeS) from the Canyon Diablo iron meteorite	CD

Key terms

Aeolian; aerosol; benthic; bivalve; Bond cycle; catastrophe theory; cellulose; chitinous exoskeleton; chlorophyll; Coleoptera; combustion-dust loading; dendrochronology; Devensian; diluvial theory; evaporite; foraminifera; General Circulation Model (GCM); glacial; glacio-isostatic rebound; gypsum; halite; Heinrich layers; Holocene; hypersaline; ice-house effect; Ice Age; ignimbrite; impact winter; interglacial; interstadial; Last Glacial Maximum; Laurentide ice sheet; Little Ice Age; loess; Messinian salinity crisis; meteorite; microfossil; Milankovitch cyclicity; North Atlantic Deep Water (NADW); nuclear winter; ozonosphere; palaeo-oceanography; palaeoclimatology; palaeoenvironmental; palaeolatitude; Pangaea; Gondwana; permafrost; pH; plankton; Pleistocene; pyrotoxin; Quaternary; radiocarbon dating; radiometric age; raised beach; redox; sapropel; sedimentary ironstone; siderophile element; solar flux; speleothem; spheroid; sporopollenin; stadial; tephra; thermohaline circulation; thermophobic; till; ultraviolet radiation; Weichselian; Wisconsin; Younger Dryas.

Issues for group discussion

Discuss the nature of climate change over the last glacial–interglacial cycle.
The discussion should focus on the results of the GRIP ice-core project (Dansgaard *et al.* 1993) and should consider the rapid changes shown by the oxygen isotope record. The discussion should also compare the relative stability in the Holocene with the variations during the last interglacial. The discussion should also consider the reasons for the differences between these interglacials.

Discuss the role of the oceanic circulation in controlling climate and climate change.

The students should examine the dynamics of oceanic circulation, particularly focusing on the thermohaline conveyor and the ENSO. The discussion should also consider the various types of proxy evidence for past changes.

Discuss the applicability of various types of proxy evidence for reconstruction of past climate.

The instructor should present different types of environments, such as high latitudes, deserts, and marine, mountain and lacustrine environments, and discuss the types and limitations of appropriate proxy data. It is useful for the students to read Lowe and Walker (1984 or 1997) and Bradley (1985).

Discuss mountain-building (orogenesis) as a cause of long-term global climate change.

Students should be encouraged to find out about current arguments over the causes of long-term (geological time) versus shorter term global climate change. Studies of seawater $^{87}Sr/^{86}Sr$ values suggest a 50–100 My episodicity, due to variations in this isotopic ratio weathered from continental rocks and/or changes in the total flux of strontium weathered from the continents (Bickle 1996). The continental weathering of strontium, as with calcium, is mainly associated with its chemical reaction with bicarbonate ions (HCO_3^{2-}). Where changes in chemical weathering rates last more than about 1 million years there may be extreme shifts in levels of atmospheric and oceanic CO_2 concentrations because the coupled ocean–atmosphere CO_2 reservoir is small, unless corresponding changes in 'solid Earth' degassing to the atmosphere can compensate. Bickle (1996) has postulated, therefore, that orogenic cycles on a *c.* 50–100 million year time scale are the most plausible cause of variations in the Sr isotope record, citing as an example the relatively rapid rise in $^{87}Sr/^{86}Sr$ values since the Miocene Period, which he links to the Himalayan orogeny. In summary, Bickle (*ibid.*) argues that long-term, geological, global climate change is strongly influenced by atmospheric CO_2 fluctuations controlled by orogenesis and chemical weathering.

Bickle, M.J. 1996. Geological controls on CO_2 degassing from the solid Earth. In: *The History of Degassing of the Earth*. Colston Research Symposium, 29–31 August 1996. Abstract volume.

Students should compare the above arguments for long-term global climate change with, for example, Milankovitch cycles as an influence. They should also be encouraged to discuss cyclic versus non-cyclic, periodic versus quasi-periodic, compared with catastrophic causes of global climate change.

Selected readings

Alvarez, L.W., Alvarez, W., Asaro, F. and Michel, H.V. 1980. An extraterrestrial impact. *Scientific American*, October, 44–60.

This paper was the first to discuss the possibility that the mass extinctions at the K–T boundary were the result of an extraterrestrial impact. It is essential for students to see this because it is a historically important paper.

Broecker, W.S. and Denton, G.H. 1990. What drives glacial cycles? *Scientific American*, 262, 42–50.

A very readable and well-illustrated review paper on the processes that may drive glacial cycles. This is a good paper to help to stimulate discussion.

Dansgaard, W. *et al.* 1993. Evidence for general instability of past climate from a 250-kyr ice core record. *Nature*, 364, 218–20.

This is an important paper presenting the work of the GRIP ice-core team. It is essential that students examine the data in this paper to aid in understanding the characteristics of Quaternary climate change.

Grootes, P.M., Stulver, M., White, J.W.C., Johnsen, S. and Jouzel, J. 1993. Comparison of oxygen isotope records from the GISP2 and GRIP Greenland ice cores. *Nature*, 366, 552–4.

This is another important paper comparing the results of the GISP2 and GRIP ice-core teams. It is an essential paper for helping to understand the use of ice-core data and as a basis for examining the nature of climate change throughout the Quaternary.

Guiot, J., Pons, A., Beaulieu, J.L. de and Reille, M. 1989. A 140,000-year continental climate reconstruction from two European pollen records. *Nature*, 338, 309–14.

This paper presents a record of climate change in Europe using pollen data. It is important for students to see this paper to examine how biological proxy data help in the reconstruction of past climate.

Rampino, M.R. and Self, S. 1992. Volcanic winter and accelerated glaciation following the Toba super-eruption. *Nature*, 359, 50–2.

A useful paper that examines the evidence that large volcanic eruptions may have been important in driving climate change. It presents a variety of proxy data, helping to illustrate their use in correlating events in different regions.

Raymo, M.E. and Ruddiman, W.F. 1992. Tectonic forcing of late Cenozoic climate. *Nature*, 359, 117–22.
This paper discusses the nature of climate change throughout the Cenozoic and attempts to relate these changes to the uplift of the Tibetan Plateau. It is a useful paper for students to examine because it considers a large variety of proxy data and discusses their limitations and the uncertainties of attributing causal factors to climate change.

Wood, B. 1992. Origin and evolution of the genus Homo. *Nature*, 355, 783–90.
This is a good review paper on the origin of humans. It is important for students to read this to assess the different arguments regarding the origins of early humans.

Textbooks

Barry, R.G. and Chorley, R.J. 1992. *Atmosphere, Weather and Climate*. London: Routledge.
This is an essential text for students and researchers who wish to acquire a strong understanding and knowledge of the dynamics and characteristics of atmospheric processes and conditions.

Bell, M. and Walker, M.J.C. 1992. Late Quaternary Environmental Change. Harlow: Longman Scientific.
This is a useful textbook on the nature of Late Quaternary environmental change. It is well written and provides a comprehensive list of references.

Bradley, R.S. 1985. *Quaternary Paleoclimatology – Methods of Paleoclimate Reconstruction*. London: Unwin Hyman.
This is a comprehensive textbook suitable for undergraduate students and researchers who wish to appreciate the various methods used in reconstructing past climate.

Bradley, R.S. and Jones, P.D. (eds) 1995. *Climate Since A.D. 1500*. London: Routledge.
This is a compilation of research work that examines the characteristics of climate change since AD 1500. It provides a good review of the nature of different types of proxy data that are being used to help to reconstruct recent palaeoclimates.

Dawson, A.G. 1992. *Ice Age Earth*. London: Routledge.
This book provides a detailed review of the fluctuations in the Earth's climate during Late Quaternary times. It is a good undergraduate text for students and researchers interested in the complex and dynamic changes that affected the Earth's surface and atmosphere during this period.

Gates, D.M. 1993. *Climate Change and its Biological Consequences*. Sutherland, Mass.: Sinauer Associates.
This is an extremely readable and well-illustrated textbook on climate change and its biological consequences. It is a useful textbook for undergraduate students who are studying the biological consequences of climate change.

Hsü, K.J. 1983. *The Mediterranean was a Desert: A Voyage of the Glomar Challenger*. Princeton, NJ: Princeton University Press.
This book provides an entertaining account of the discovery of the evidence that led to the proposal that the Mediterranean Sea had evaporated. It is somewhat outdated, but it provides the reader with an insight into the scientific work and methodologies of Earth scientists.

Imbrie, J. and Imbrie, K.P. 1979. *Ice Ages: Solving the Mystery*. London: Macmillan.
This very readable text provides an account of the causes and effects of Ice Ages. Although somewhat dated, it is strongly recommended to any student and teacher who wants a good historical background in global climate change.

Lamb, H.H. 1995. *Climate History and the Modern World*, second edition. London: Routledge.
This is a thought-provoking book that examines the links between civilisation and climate change throughout history. It is essential background reading for any student studying palaeoclimatology.

Lowe. J.J. and Walker, M.J.C. 1997. *Reconstructing Quaternary Environments*, second edition. Harlow: Longman.
This is an important text for students and professionals concerned with Quaternary Science. It examines the various forms of evidence that can be used to establish the history and scale of environmental change through Quaternary time. New ideas and methods are presented in an interesting manner. Furthermore, the text is well written and illustrated, and has an up-to-date and comprehensive bibliography.

Williams, M.A.J., Dunkerley, D.L., Deckker, P. De, Kershaw, A.P. and Stokes, T. 1993. *Quaternary Environments*. London: Edward Arnold.
This is a comprehensive and well-illustrated text that examines the environmental changes that have taken place throughout Quaternary times. Besides being a useful undergraduate text for students studying Quaternary environmental change it is also a useful source for references.

Essay questions

1 'In looking for evidence of climatic impact in the course of history, it is sensible to look most at the marginal areas near the poleward and arid limits of human settlement and activity, for it is there that vulnerability is likely to be greatest.' (Lamb, 1995, p. 3). Discuss.

2 Describe the natural processes that may result in global climate change.

3 Describe the different types of proxy data that are used to reconstruct past climates, and discuss any limitations associated with such data.

4 Why is the study of Quaternary climatic change important for an understanding of human-induced climate change?

5 Describe examples of global climate change throughout the geological record, and explain how such changes may have come about.

6 Discuss the proposition that just as politicians should study history, so too should those who wish to understand present and future climatic change study the geological record.

7 Examine the assertion that climatic changes are characterised by periods of abrupt and rapid change.

8 Discuss the role of oceanic circulation in controlling global climate.

9 Outline and discuss the main environmental changes that might be expected to occur through a Quaternary glacial–interglacial cycle in a cool-temperate region such as the British Isles.

10 With reference to the last glacial cycle, discuss the scales and magnitudes of environmental change.

11 Describe and evaluate briefly the types of evidence used to reconstruct environmental change during the last c. 3 million years.

12 Discuss the ways in which EITHER the physical environment OR the biota of the British Isles or the USA would respond to a drop in mean annual temperature of approximately 10 °C.

Multiple-choice questions

Choose the best answer for each of the following questions.

1 The depth in the ocean where sea water is under saturated with respect to $CaCO_3$ is known as the:
  (a) calcium compensation depth
  (b) calcium compensation zone
  (c) carbonate compensation zone
  (d) carbonate compensation depth

2 The hypothesis that important and widespread changes in the physical environment are brought about by major, relatively brief, and sudden events is known as:
  (a) Darwinian theory
  (b) catastrophe theory
  (c) calamity theory
  (d) geomorphic theory

3 Most Quaternary scientists consider the start of the Quaternary Period to be approximately:
  (a) 2.5 Ma
  (b) 1.64 Ma
  (c) 140 ka
  (d) 10 ka

4 A global sudden cold period between about 11,000 and 10,500 years BP is known as the:
  (a) Younger Dryas
  (b) Holocene
  (c) Eemian
  (d) Little Ice Age

5 One of the most commonly used names for the last glacial is the:
  (a) Devensian
  (b) Wisconsin
  (c) Weichselian
  (d) a, b and c

6 Which of the following isotopic ratios is most used to help reconstruct former sea-surface temperatures and sea levels:
  (a) $^{18}O/^{16}O$
  (b) $^{18}O/^{14}O$
  (c) $^{87}Sr/^{86}Sr$
  (d) $^{87}Sr/^{84}Sr$

7 The Last Glacial Maximum (LGM) occurred approximately:
  (a) 5,000 years BP
  (b) 80,000 years BP
  (c) 18,000 years BP
  (d) 22,000 years BP

8 Milankovitch cycles are produced by fluctuations in:
(a) sunspot activity
(b) volcanic activity
(c) the volume of ice sheets
(d) the Earth's orbital parameters

9 Approximately 5 Ma the waters of the Mediterranean almost dried up, which increased the salinity of the waters and led to the precipitation of large salt deposits. This episode is known as the:
(a) Mediterranean salinity crisis
(b) Messinian salinity crisis
(c) Wisconsin salinity crisis
(d) Pliocene salinity crisis

10 The most probable cause of the Little Ice Age was:
(a) sunspot activity
(b) volcanic activity
(c) glacier fluctuations
(d) human emission of greenhouse gases

11 The best fossils for reconstructing past climates are:
(a) diatoms
(b) coleoptera
(c) foraminifera
(d) pollen

12 Dendrochronologies are time correlations based on the use of:
(a) ice layers in ice sheets
(b) volcanic ash layers
(c) moraines
(d) tree rings

13 Which is the correct term for a cold period which has a duration of between about 100 and 1,000 years?
(a) glacial
(b) stadial
(c) interstadial
(d) Ice Age

14 Heinrich layers are:
(a) very thin coarse-grained layers of deep-marine sediments
(b) fossil soils within loess
(c) layers of volcanic ash within glacial ice
(d) thin tree rings which had been affected by severe frost

15 Rocks at the Cretaceous–Tertiary boundary contain high concentrations of:
(a) iridium
(b) diamonds
(c) charcoal
(d) a, b and c

16 The biggest mass extinction occurred at the:
(a) Cretaceous–Tertiary boundary
(b) end of the Permian Period
(c) end of the last glacial
(d) Ordovician–Silurian boundary

17 The glacial theory was presented by:
(a) James Hutton
(b) Louise Agassiz
(c) Milutin Milankovitch
(d) Reverend William Buckland

18 The oldest *Homo* is:
(a) *Homo habilis*
(b) *Homo erectus*
(c) *Homo sapiens neanderthalensis*
(d) *Homo sapiens sapiens*

19 The best continental record of climate change over the whole of the Quaternary Period is provided by:
(a) the Antarctic Ice Sheet
(b) the Greenland Ice Sheet
(c) the Chinese loess
(d) the Skjonghelleren cave sediments

20 The Quaternary Ice Age was probably initiated by:
(a) tectonic factors
(b) changes in the Earth's orbital parameters
(c) biological changes
(d) sunspot activity

21 Approximately 10,000 years ago global sea levels were approximately:
(a) 10 m higher
(b) 10 m lower
(c) 30 m lower
(d) 150 m lower

22 Which of the following volcanic eruptions was the largest during Quaternary times?
(a) Toba (*c.* 75 ka)
(b) Laacher See (*c.* 11 ka)
(c) Krakatau (1883)
(d) Pinatubo (1991)

23 Which of the following isotopes are the most appropriate to use to calculate changes in biomass productivity?
   (a) ^{18}O and ^{16}O
   (b) ^{12}C and ^{13}C
   (c) ^{15}N and ^{14}N
   (d) ^{87}Sr and ^{86}Sr

24 The Laurentide Ice Sheet covered:
   (a) most of Canada
   (b) all of North America
   (c) North America and Europe
   (d) Europe

25 Modern humans spread to Australia about:
   (a) 50,000 years ago
   (b) 18,000 years ago
   (c) 10,000 years ago
   (d) 8,000 years ago

Figure questions

1 Figure 2.2 is a generalised global oceanic circulation. Answer the following questions.
   (a) What is this oceanic circulation called?
   (b) Why is this oceanic circulation important for understanding global climate?
   (c) What are the likely effects if this oceanic circulation ceased to operate?

2 Figure 2.12 illustrates the variability in the Earth's orbit around the Sun at various time scales. Answer the following questions.
   (a) What combined effect do these orbital parameters have on global climate?
   (b) Which of the orbital parameters has the strongest effect on climate?
   (c) Draw schematic graphs to show the power spectra for eccentricity, obliquity and precession, and their combined effect.

3 Figure 2.25 illustrates the changes in $\delta^{13}C$ (‰), $\delta^{18}O$ (‰) and $\delta^{87}Sr/^{86}Sr$ isotopes from marine sediments throughout Cenozoic times. Answer the following questions.
   (a) Using the appropriate graph(s), describe the characteristic of climate change during Cenozoic times.
   (b) In terms of uplift, what do the $\delta^{87}Sr/^{86}Sr$ isotopes show?
   (c) What do the $\delta^{13}C$ isotopes indicate?

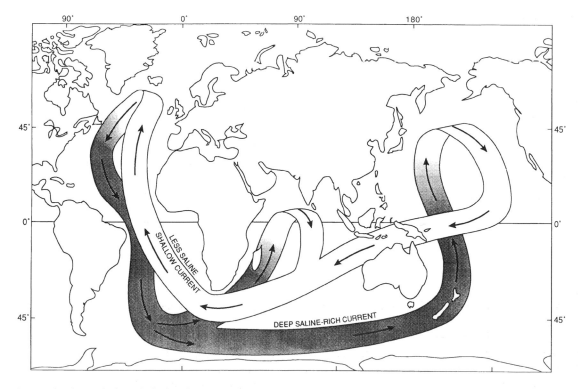

Figure 2.2 *The thermohaline (salt–heat) conveyor belt in the oceans. Solid arrows show the flow of deep, cold and salty water; open arrows show the return flow. Notice that the deep currents begin in the North Atlantic, in the East Greenland Sea, then move southwards from the Atlantic into the Pacific Ocean. The upper, warmer current may begin in the tropical seas around Indonesia, and includes the strong flow out of the Gulf of Mexico. Redrawn after Street-Perrott and Perrott (1990).*

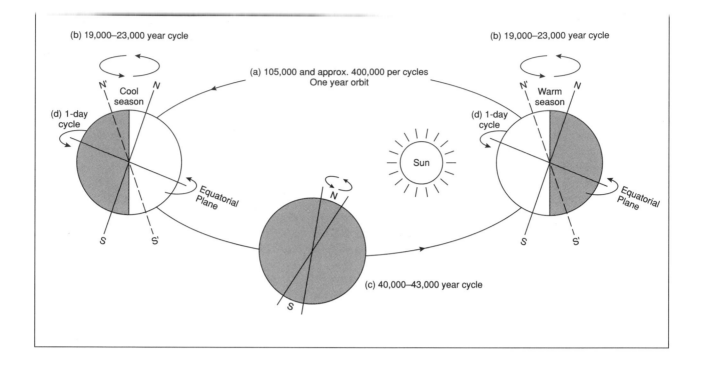

Figure 2.12   *The variability in the Earth's orbit around the Sun at various time scales measured in tens of thousands of years, and referred to as orbital parameters. The temporal variation in orbital parameters causes long-term changes in the amount of solar energy reaching the surface of the Earth, which in turn can result in significant changes in global climate, referred to as Milankovitch cyclicity, so named after one of the first people to propose a link between changes in the Earth's orbit and global climate change. Adapted from Peltier (1990).*

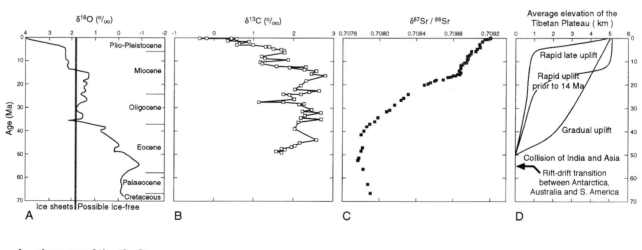

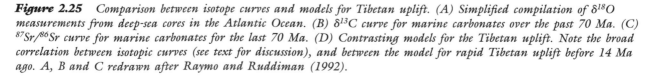

A = deep-sea Atlantic Ocean cores
B and C = marine carbonates
NB. broad correlation with uplift prior to 14 Ma

Figure 2.25   *Comparison between isotope curves and models for Tibetan uplift. (A) Simplified compilation of $\delta^{18}O$ measurements from deep-sea cores in the Atlantic Ocean. (B) $\delta^{13}C$ curve for marine carbonates over the past 70 Ma. (C) $^{87}Sr/^{86}Sr$ curve for marine carbonates for the last 70 Ma. (D) Contrasting models for the Tibetan uplift. Note the broad correlation between isotopic curves (see text for discussion), and between the model for rapid Tibetan uplift before 14 Ma ago. A, B and C redrawn after Raymo and Ruddiman (1992).*

Short questions

1 How is the base of the Quaternary Period defined?

2 What are sunspots?

3 Describe the dynamics of an El Niño event.

4 List ten different types of proxies that can be used to help to reconstruct past climate.

5 Outline the basic characteristic of climate change throughout the Quaternary Period.

6 What is Milankovitch cyclicity?

7 Describe the likely consequences of a large meteorite colliding with the Earth.

8 List the main extinctions that have occurred throughout geological time.

9 Describe what happened during the Messinian salinity crisis.

10 Why are the thick loess successions in China important for the study of Quaternary palaeoclimatology?

Answers to multiple-choice questions

1 d; 2 b; 3 a; 4 a; 5 d; 6 a; 7 c; 8 d; 9 b; 10 a; 11 b; 12 d; 13 b; 14 a; 15 d; 16 b; 17 b; 18 a; 19 c; 20 a; 21 c; 22 a; 23 b; 24 a; 25 a.

Answers to figure questions

1 (a) Thermohaline conveyor. (b) Oceanic circulation is important for understanding global climate because it transfers vast amounts of heat energy throughout the globe. (c) If the thermohaline conveyor ceased to operate it would drastically alter global heat balances and it is likely that cold waters travelling to warmer latitudes would not warm up. This would lead to increased cooling in high-latitude regions, ultimately leading to global cooling.

2 (a) Combined, the orbital parameters control the temporal and spatial variations in the distribution of solar energy over the Earth, which ultimately controls natural climate change. (b) Although the variation in the incoming solar radiation is relatively weak for eccentricity compared with the precession and obliquity periods, the 100-kyr signal appears to dominate. (c) See Figure 2.12.

3 (a) Graph A shows the variation of $\delta^{18}O$ values throughout the last 70 Ma. This shows that there has been a general cooling trend since early Eocene times, punctuated by abrupt cooling trends. The most notable abrupt changes are at the boundary of the Eocene and Oligocene Periods and at about 2–5 Ma with the onset of the Quaternary Ice Age. (b) $\delta^{87}Sr/^{86}Sr$ isotopes are used as a proxy for erosion rates. The increased ratios suggest that erosion of the Himalayas was accelerating from about 40 Ma, suggesting that the Himalayas were uplifting. After about 15 Ma, the rate of change decreased, suggesting that the Himalayas may have reached their present elevations soon after about 15 Ma. The $\delta^{87}Sr/^{86}Sr$ isotopes, however, can be interpreted by other processes, such as increased Sr input into the oceans by erosion as the great ice sheets began to grow. (c) The $\delta^{13}C$ isotopes are a proxy for bioproductivity and carbon storage in the oceans. Decreasing values indicate that carbon burial decreased throughout Tertiary times; this in turn could have caused additional CO_2 to be released into the oceans and atmosphere. Since oxygen is necessary to oxidise organic matter and release CO_2, the decrease in buried organic carbon may reflect an increase in dissolved oxygen concentrations in the oceans. Oxygen solubility increases with decreasing temperature, therefore the dissolution of oxygen into the oceans would have increased as the oceans began to cool, thus partially explaining this trend.

Answers to short questions

1 The *International Commission on Stratigraphy* formally defines the base of the Quaternary Period where a claystone horizon containing the first appearance of a cold-loving (thermophobic) microfossil (*foraminifera*) directly overlies a black mud rich in organic calcium carbonate (called a sapropel). Most Quaternary scientists prefer, however, to define the base as being much older. They extend it back to about 2.5 Ma, when the oxygen isotopic record shows that the world's climate became very cold and the Earth entered the present Ice Age.

2 Sunspots are areas of cooler gas and stronger magnetic fields in the Sun's surface or photosphere. Typically, the observed temperature of a sunspot is

about 3,900 K, compared with the background normal photosphere at 5,600 K.

3 El Niño events are relatively large perturbations of a climatic process that occurs annually in the Pacific Ocean. During an El Niño event the sea-surface temperature in the tropical Pacific Ocean reaches a minimum 0.5 °C warmer than normal for at least six consecutive months. The underlying cause of El Niño events is the eastward propagation of a down-welling Kelvin wave across the equatorial Pacific Ocean. The eastward-propagating Kelvin waves are confined to a narrow belt by the Coriolis force. In a 'normal year', the variations in the atmosphere–ocean system produce a fairly predictable pattern of ocean currents in the southern Pacific Ocean. The sea-surface temperature is highest in the West (>28 °C), which helps to induce the movement of strong, warm, maritime southeast trade winds into Indonesia and with them heavy rainfall. A corollary of this is that cold, nutrient-rich bottom water upwells to replenish surface waters off the western coast of South America. In contrast to such normal years, during an El Niño, surface waters greater than 28 °C develop much farther eastwards. This allows the intertropical convergence zone to migrate southwards and suppress the southeast trades, or even reverse them, as the rainfall is heaviest in the central-east Pacific and upwelling of cold, nutrient-rich bottom water is weaker.

4 Different types of proxies for climate change include rock types; fossils; stable isotopes; nitrogen isotopes; tree rings; geomorphology; historical records; heavy minerals; trace metals; and pH of ice cores

5 The climate during Quaternary times, from about 2.5 Ma to the present, was colder than in the previous geological times, i.e. Tertiary times. The climate fluctuated from predominantly cold times (glacials) lasting between 100,000 and 200,000 years to warm periods lasting 10,000 to 20,000 years. Small changes in climate were superimposed on the climates of glacials and interglacials. These are known as stadials (cold) and interstadials. Changes from warm to cold, and vice versa, probably occurred very rapidly, in the order of a few decades to hundreds of years.

6 Milankovitch cyclicity, named after the astronomer of the same name, is the slight changes in the solar flux to the Earth's surface caused by variations in the Earth's orbit. The changes in climate from glacials to interglacials are probably driven by Milankovitch forcing.

7 If a large meteorite impacts the Earth, such as is likely happened at the Cretaceous–Tertiary boundary event, the consequences may be catastrophic. The impact may cause global fires, enhance levels of atmospheric aerosols and reduce sunlight, which in turn lead to global cooling, and poisonous chemicals having extremely serious effects on life forms. Other effects may include very acidic rain, a depletion of the stratospheric ozone layer, and decalcification of the oceans.

8 Five major mass extinctions have occurred throughout geological time. These occurred at about the boundary of the Ordovician and Silurian Periods, when there was a major glaciation; 375 Ma, late in the Devonian Period; 240 Ma, at the boundary between the Permian and Triassic Periods, when 80–95 per cent of organisms became extinct; 210 Ma, in the Triassic Period; and at 65 Ma, at the boundary between the Cretaceous and Tertiary Periods, which was probably the result of meteorite(s) colliding with the Earth and resulted in the extinction of the dinosaurs.

9 The Messinian salinity crisis refers to the events that occurred about 5–6 million years ago. During this time, the Mediterranean Sea became landlocked as a result of plate tectonic processes that uplifted the Straits of Gibraltar, sealing off the waters of the Atlantic Ocean from the Mediterranean Sea. The landlocked Mediterranean Sea evaporated and changed the regional climate to desert conditions. The evaporation of the sea waters, probably periodically replenished by catastrophic flooding from the Atlantic Ocean, caused the accumulation locally of up to 1 km in thickness of salts. About 5 million years ago, the Straits of Gibraltar were breached by the Atlantic Ocean waters, which then flooded back into the Mediterranean.

10 The thick loess successions in China are important because they provide the most continuous and detailed record for continental climate change throughout Quaternary times. The loess reaches a maximum thickness of 330 m in China, and small changes in grain size, mineralogy, magnetic susceptibility and palaeontology record subtle changes in climate. As many as 32 palaeosols are present at some locations. These palaeosols reflect periods of increased warmth and humidity. Sedimentological characteristics of the loess succession correlate well with the marine oxygen isotopic record, which suggests that climate change on the continents correlates well with the changes predicted by Milankovitch theory.

Additional references

Barbeau, K., Moffett, J.W., Caron, D.A., Croot, P.L. and Erdner, D.L. 1996. Role of protozoan grazing in relieving iron limitation of phytoplankton. *Nature*, 380, 61–4.

Bard, E., Hamelin, B., Arnold, M., Montaggioni, L., Cabioch, G., Faure, G. and Rougerie, F. 1996. Deglacial sea-level record from Tahiti corals and the timing of global meltwater discharge. *Nature*, 382, 241–4.

Behl, R.J. and Kennett, J.P. 1996. Brief interstadial events in the Santa Barbara basin, NE Pacific, during the past 60 kyr. *Nature*, 379, 243–6.

In this paper, the authors identify Dansgaard–Oeschger (D–O) climatic cycles in the eastern equatorial Pacific, thereby further extending their identification from the Greenland ice cores and the North Atlantic. During the past 100 kyr, the Santa Barbara basin laminated sediments preserve at least 20 warm (interstadial) D–O events.

Colman, S.J., Peck, J.A., Karabanov, E.B., Carter, S.J., Bradbury, J.P., King, J.W. and Williams, D.F. 1996. Continental climate response to orbital forcing from biogenic silica records in Lake Baikal. *Nature*, 378, 769–71.

de Vernal, A., Hillaire-Marcel, C. and Bilodeau, G. 1996. Reduced meltwater outflow from the Laurentide ice margin during the Younger Dryas. *Nature*, 381, 774–7.

DeWeaver, E. and Nigam, S. 1995. Influence of mountain ranges on the mid-latitude atmospheric response to El Niño events. *Nature*, 378, 706–8.

During El Niño events, mountain belts are shown to exert a significant influence on atmospheric vorticity (a measure of local air circulation about a vertical axis) in the upper troposphere extending into mid-latitude regions. When a rotating air mass encounters a mountain range, it is compressed vertically, spreads horizontally and its total vorticity is reduced. The net result is a significant change in weather over and in the vicinity of the mountain range.

Farley, K.A. and Patterson, D.B. 1995. A 100-kyr periodicity in the flux of extraterrestrial ^{3}He to the sea floor. *Nature*, 378, 600–3.

Since most of the ^{3}He in oceanic sediments comes from interplanetary dust particles (IDPs), a measure of its temporal variability in slowly accumulating pelagic clays in the oceans might be expected to show a Milankovitch cyclicity. Farley and Patterson demonstrate a 100-kyr periodicity in Quaternary sediments from the Mid-Atlantic Ridge that may well be related to the 100-kyr eccentricity in the Earth's orbit.

Field, M.H., Huntley, B. and Müller, H. 1994. Eemian climate fluctuations observed in a European pollen record. *Nature*, 371, 779–83.

Gallimore, R.G. and Kutzbach, J.E. 1996. Role of orbitally induced changes in tundra area in the onset of glaciation. *Nature*, 381, 503–5.

Kapista, A.P., Ridley, J.K., Robin, G. de Q., Siegert, M.J. and Zotikov, I.A. 1996. A large, deep, freshwater lake beneath the ice of central East Antarctica. *Nature*, 381, 684–6.

Kuniholm, P.I., Kromer, B., Manning, S.W., Newton, M., Latini, C.E. and Bruce, M.J. 1996. Anatolian tree rings and the absolute chronology of the eastern Mediterranean, 2220–718 BC. *Nature*, 381, 780–3.

Kutzbach, J., Bonan, G., Foley, J. and Harrison, S.P. 1996. Vegetation and soil feedbacks on the response of the African monsoon to orbital forcing in the early to middle Holocene. *Nature*, 384, 623–6.

Fossil pollen, ancient lacustrine sediments and archaeological evidence are used to show that during the early to middle Holocene Period (c. 12,000–5,000 years BP) the Sahel and Sahara regions were considerably wetter than at present.

Lao, Y., Anderson, R.F., Broecker, W.S., Trumbore, S.E., Hofmann, H.J. and Wolfli, W. 1992. Increased production of cosmogenic ^{10}Be during the Last Glacial Maximum. *Nature*, 357, 576–8.

Beryllium-10 (^{10}Be), which has a half-life of 1.5×10^6 years, is produced in the upper atmosphere during the bombardment of nitrogen and oxygen atoms by cosmic rays. This nuclear process is influenced by the strength of the Earth's magnetic field and solar activity, e.g. sunspot cycles. At times when the Earth's magnetic field is relatively weak, more cosmic rays can bombard the upper atmosphere, increasing the production of ^{10}Be and, therefore, its flux to the oceans and land surface. In this paper, Lao *et al.* report a study of Pacific Ocean deep-sea sediments where the global average production rate of ^{10}Be was at least 25 per cent greater around the Last Glacial Maximum (*c.* 24,000–16,000 years ago) than in the past 10,000 years (Holocene). There are data to suggest that about 23,000–15,000 years BP the Earth's magnetic field may have been 50 per cent less than during the past 10,000 years (Mazaud *et al.* 1991). Not only is this study interesting on its own, but the ^{10}Be results are consistent with the similar arguments used to account for the younger ^{14}C ages of fossil corals compared with age dates obtained by U–Th methods.

Linsley, B.K. 1996. Oxygen-isotope record of sea level and climate variations in the Sulu Sea over the past 150,000 years. *Nature*, 380, 234–7.

The Sulu Sea, in the western Pacific, has the highest mean annual sea-water temperatures anywhere on Earth and, therefore, is a major contributor of water vapour to the atmosphere. This paper examines the δ^{18}O record of planktonic foraminifera extending back to 150,000 years BP, and shows that during Interglacial Stage 3 (58,000–23,000 years BP), when the Sulu Sea record deviates considerably from the SPECMAP deep-sea δ^{18}O record, sea level was only 40–50 m below present level, i.e. higher than previously inferred solely from the deep-sea record.

Mann, M.E., Park, J. and Bradley, R.S. 1995. Global interdecadal and century-scale climate oscillations during the past five centuries. *Nature*, 378, 266–70.

Using data obtained from a multivariate analysis of a globally distributed set of temperature proxy data records spanning several centuries, Mann *et al.* suggest that interdecadal (15–35 year period) and century-scale (50–150 year period) climate variability may be intrinsic to the natural climate system.

Mazaud, A., Laj, C., Bard, E., Arnold, M. and Tric. E. 1991. *Geophysical Research Letters*, 18, 1885–8.

Overpeck, J., Rind, D., Lacis, A. and Healy, R. 1996. Possible role of dust-induced regional warming in abrupt climate change during the last glacial period. *Nature*, 384, 447–9.

Ramaswamy, V. and Nair, R. 1992. Measuring the monsoon. *New Scientist*, 20 June, 33–5.
This article provides a useful introduction to the nature of the monsoons, and how they affect the delivery rate of sediments to the deep sea. It also suggests that the monsoons have existed since *c.* 12 Ma. It appears that although dams are now trapping much of the natural monsoon-related terrigenous sediment flux to the deep sea, the phytoplankton and zooplankton blooms produce biomass that traps river-derived (monsoon-flushed) pollutants and fine sediment particles in 'mats' that would otherwise sink much more slowly to the sea bed. These particles sink as a marine snow much faster than individual particles, so that sediments in the deep sea can record annual events, like monsoons, only weeks to months after any surface-water event.

Weaver, A.J. and Hughes, T.M.C. 1994. Rapid interglacial climate fluctuations driven by North Atlantic Ocean circulation. *Nature*, 367, 447–50.
This paper explores the computer modelling of three simulated North Atlantic Ocean states: 'warmer than', 'similar to' and 'colder than' today, all conditions that appear to have characterised the Eemian interglacial compared with the present interglacial (Holocene). Lao *et al.* show that by the introduction of a random component to the mean fresh-water flux, which seems to drive the oceanic conveyor belt, rapid (climatic) fluctuations between the three states occur.

Aims

- To evaluate the evidence for human-induced climate change.

- To examine the atmospheric processes that control climate and that are being affected by human activity.

- To assess the environmental, economic, social and political consequences of global atmospheric change.

- To examine methods to reduce the possible effects of human-induced global atmospheric change.

Key point summary

- Although there have been rapid fluctuations in climate during and between Ice Ages, the Earth's self-regulation mechanism, particularly through the **carbonate–silicate geochemical cycle**, has kept the climate stable for 3,800 million years, even though the solar flux has increased. However, human activity is threatening the stability of this climate.

- Human activity is affecting the atmosphere in a number of ways. The main ones include changes in the level of **stratospheric** and **tropospheric ozone**; increases in the concentration of **greenhouse gases**; and increases in the prevalence of **smog**.

- Stratospheric ozone (O_3) forms part of an important natural shield around the Earth that absorbs large amounts of **UV-B radiation** from the Sun.

- Significant depletion of O_3 concentrations has been recorded since 1977 and are now common over the poles following Antarctic spring, when supercooled clouds act as a catalyst for heterogeneous reactions that destroy the ozone.

- Stratospheric O_3 depletion, at least in part, is the result of anthropogenic emissions of **CFCs** and other chemicals, as well as natural processes, such as volcanic eruptions, that produce aerosols and acids.

- Reduced stratospheric O_3 levels result in a cooling of the stratosphere and warming of the troposphere; the net result is a cooling of the Earth's surface. Latitude, altitude and seasons strongly influence this balance.

- Increased atmospheric CO_2 could lead to tropospheric warming and is likely to result in increased cloud cover, which may activate O_3-depleting species and thereby increase the rate of O_3 depletion.

- Tropospheric ozone concentrations, however, have been increasing in polluted areas, where it is an important greenhouse gas and contributes to poisonous photochemical smog.

- The **greenhouse effect** warms the Earth's atmosphere. **Short-wavelength (ultraviolet) solar radiation** reaches the Earth's surface and is re-radiated as **long-wavelength (infrared) radiation** back up into the atmosphere, where greenhouse gases absorb it and warm the Earth.

- The most important greenhouse gases include water vapour; carbon dioxide (CO_2); methane (CH_4); tropospheric ozone (O_3); nitrous oxide (N_2O); ammonia (NH_3); CFCs; and halons. These have different **global warming potentials** (GWPs) and varying **residence times** in the atmosphere.

- An increase in greenhouse gases may lead to global warming. However, there are many negative feedbacks that counteract any potential warming.

- **General Circulation Models (GCMs)** are a means of assessing the climatic sensitivity to changing atmospheric concentrations of greenhouse gases. The Intergovernmental Panel on Climate Change (IPCC) assessment of these suggests that a doubling of atmospheric CO_2 concentrations is likely to increase the global mean surface temperature by between 1.5 °C and 4.5 °C. If greenhouse gas emissions continue, there could be an increased severity of storms and droughts. In addition there could be an overall rise in sea level of 50–60 cm by the year 2100. This is due to the thermal expansion of the oceans and melting of ice caps and glaciers.

- Evidence for human-induced global warming is equivocal because of the difficulties of comparing historical and modern meteorological data. In addition, it is not known if changes are due to natural perturbations in the Earth's atmosphere and hydrosphere or to anthropogenic causes.

- Natural phenomena also cause global climate change, e.g. **volcanic eruptions** and **El Niño events**.

- Volcanic eruptions emit huge quantities of aerosols, which reduce incoming solar radiation and lead to global cooling of 0.2–0.5 °C for short durations (<5 years). Volcanic activity also emits greenhouse gases (CO_2), ozone-depleting gases (Cl_2, F_2) and gases that may act as catalysts for ozone-depleting reactions (SO_2, which forms H_2SO_4). The gases cause either negative or positive feedback mechanisms for climate change.

- El Niño events produce relatively small changes in the Earth's climate, due to variations in the atmosphere–ocean system. Variations in oceanic circulation and upwelling of nutrient-rich, deep-ocean currents affect biomass production (e.g. algal blooms, etc.) and atmospheric CO_2 concentrations. During an El Niño event, more CO_2 is released into the atmosphere and heavy rainfall occurs in the southern Pacific Ocean. This causes droughts in Indonesia and Australia and floods in South America.

- International action on global atmospheric pollution has resulted in attempts to reduce ozone depletion (Montreal Protocol, 1987; with amendments in Helsinki, 1989, and Copenhagen, 1992) and reductions in greenhouse gas emissions (UN Framework Convention on Climate Change, 1992).

Main learning hurdles

Chemical formulae

Students with poor science backgrounds are easily deterred by the use of scientific notation. Although this chapter does not use complex equations, the atmospheric gases are referred to by their chemical formulae. The instructor should explain these.

Reduction–oxidation (redox) reactions

The principal global geochemical cycle of redox energy can be considered as the photochemical reactions involving the reaction of water (H_2O) and carbon dioxide (CO_2) to form free oxygen (O_2) and organic compounds (the latter being considered in its simplest form as CH_2O), which then react with each other and/or various chemical species, including ions. A by-product of these reactions is the production of water and carbon dioxide that can then react with each other to continue the cycle.

In summary, the main *oxidation reactions* are as follows:

$$O_2 + CH_2O \rightarrow CO_2 + H_2O$$
$$O_2 + N_2 \rightarrow NO_3^- + H_2O$$
$$O_2 + H_2S \rightarrow SO_4^{2-} + H_2O$$
$$O_2 + CH_4 \rightarrow CO_2 + H_2O$$

and the main *reduction reactions* are:

$$CH_2O + O_2 \rightarrow H_2O + CO_2$$
$$CH_2O + NO_3^- \rightarrow N_2 + CO_2$$
$$CH_2O + SO_4^{2-} \rightarrow H_2S + CO_2$$
$$CH_2O + CO_2 \rightarrow CH_4 + CO_2$$

Radiation balances

Many students do not understand the physics of radiation and heat exchange. The instructor must, therefore, explain the nature of electromagnetic radiation and the various ways that radiation flows within the Earth's atmosphere and how it contributes to global heat exchange.

Feedback systems

The instructor should revise the section on systems and feedbacks that was discussed in Chapter 1. The student will then be able to appreciate fully the terminology and relevance of climatic feedbacks. This is particularly important when assessing possible future changes due to anthropogenic emissions of greenhouse gases and ozone depletion.

Key terms

CFCs; El Niño Southern Oscillation (ENSO); General Circulation Model (GCM); global warming potential (GWP); halocarbon; heat-island effect; hydroxyl radical; infrared radiation; metamorphism; methane; nitrous oxide; photochemical reaction;

stratospheric ozone; thermocline; tropospheric ozone; ventilation.

Issues for group discussion

Discuss the likely changes in weather conditions that may occur due to global warming, and the economic, political and social implications.
The students must read and discuss Golnaraghi *et al.* (1995), Hill (1995), Kerr (1995) and Parry and Swaminathan (1992). Emphasis must include the uncertainties involved in predicting such changes.

Discuss the problems associated with attempting to implement carbon taxes.
The students should be encouraged to discuss the problems of imposing carbon taxes on developing countries and other mechanisms that may help to reduce CO_2 emissions.

Examine the problems of urban air pollution.
It would be useful for students to read Lee and Manning (1995), Lents and Kelly (1993), UNEP and WHO (1994), and Stone (1995) before the discussion. Students should emphasise the environmental health problems and the reduction in the quality of life.

Selected readings

Chen, D., Zebiak, S.E., Busalacchi, A.J. and Cane, M.A. 1995. An improved procedure for El Niño forecasting: implications for predictability. *Science*, 269, 1699–702.
This is a very useful paper discussing the characteristics of El Niño and the possibility of future forecasting.

Deming, D. 1995. Climatic warming in North America: analysis of borehole temperatures. *Science*, 268, 1576–7.
This paper describes a useful technique for assessing recent temperature changes in North America. It provides a good basis from which to start discussions of the significance of data and the likelihood of global warming.

Golnaraghi, M. and Kaul, R. 1995. Responding to ENSO. *Environment*, 37 (1), 16–20 and 38–44.
This is a good review of the processes and consequences of El Niño events. It provides a wide perspective and is useful for discussing the role of climate change on human activity.

Hill, D.K. 1995. Pacific warming unsettles ecosystems. *Science*, 267, 1911–12.
A very interesting short paper examining one of the many effects on ecosystems that may result from global warming. It provides a basis from which students can start discussing the likely consequences of global warming.

Kerr, R.A. 1995. US climate tilts towards the greenhouse. *Science*, 268, 363–4.
An interesting comment on recent weather conditions in the USA and the possibility that they may be a forewarning of the likely weather that may occur due to global warming.

Lee, J. and Manning, L. 1995. Environmental lung disease. *New Scientist*, 16 September, Inside Science 84.
This is a useful discussion of the nature of environmental lung diseases.

Lents, J.M. and Kelly, W.J. 1993. Clearing the air in Los Angeles. *Scientific American*, 269, 18–25.
This paper discusses the methods that help to reduce the problems of urban air pollution. It provides a useful example that may apply to other regions.

McCormick, M.P., Thomason, L.W. and Trepte, C.R. 1995. Atmospheric effects of the Mt Pinatubo eruption. *Nature*, 373, 399–404.
This well-illustrated review paper describes the climatic effects of the eruption of Mount Pinatubo.

Parry, M.L. and Swaminathan, M.S. 1992. Effects of climatic change on food production. In: Mintzer, M.I. (ed.), *Confronting Climatic Change: Risks, Implications and Responses*, 113–27. Cambridge: Cambridge University Press.
This is a good review of the likely consequences of climate change on food production and provides a useful basis for considering the economic, political and social implications of climate change.

Plaut, G., Ghil, M. and Vautard, R. 1995. Interannual and interdecadal variability in 335 years of central England temperatures. *Science*, 268, 710–13.
A useful paper presenting data on the variability of temperature in central England that help to illustrate the nature of climate change over the last 300 years.

United Nations Environmental Programme and the World Health Organisation 1994. Air pollution in the world's megacities. *Environment*, 36 (2), 4–13 and 25–37.
This is an important report summarising the problems and nature of urban air pollution. It provides a useful

database to consider the consequences of pollution on urban populations.

Stone, R. 1995. If the mercury soars, so may health hazards. *Science*, 267, 957–8.
An interesting note considering the possible health problems in urban environments if global warming occurs.

Watson, A. 1991. Carbon dioxide. *New Scientist*, 6 July, Inside Science, 48.
This is a useful description of the importance of carbon dioxide in the carbon cycle.

Wuethrich, B. 1995. El Niño goes critical. *New Scientist*, 4 February, 33–5.
This is an interesting article discussing the increased frequency of El Niño events during the last few years. It is useful because it addresses the problems and uncertainties of attributing a forcing mechanism to El Niño events.

Textbooks

Boyle, S. and Ardill, J. 1989. *The Greenhouse Effect*. London: Hodder & Stoughton.
Although a little dated this is a highly readable text useful for the lay-person with little scientific background or the social scientist.

Graedel, T.E. and Crutzen, P.J. 1993. *Atmospheric Change: An Earth System Perspective*. New York: W.H. Freeman & Co.
This book provides a useful introduction to atmospheric chemical processes for the more science-based students.

Houghton, J.T., Jenkins, G.J. and Ephraums, J.J. (eds) 1990. *Climate Change: The IPCC Scientific Assessment*. Cambridge: Cambridge University Press.
Houghton, J.T., Callander, B.A. and Varney, S.K. (eds) 1992. *Climate Change 1992: The Supplementary Report to the IPCC Scientific Assessment*. Cambridge: Cambridge University Press.
Houghton, J.T., Meira Filho, L.G., Hoesung Lee, Callander, B.A., Haites, E., Harris, N. and Maskell, K. (eds) (Intergovernmental Panel on Climate Change) 1995. *Climate Change 1994. Radiative Forcing of Climate Change and an Evaluation of the IPCC IS92 Emission Scenario*. Cambridge: Cambridge University Press.
These three reports are the result of Working Group 1 of the Intergovernmental Panel on Climate Change, set up by the World Meteorological Organisation and the United Nations Environment Programme. They are essential reading and reference material for anyone interested in global climate change. The reports assess the potential effects that human activity may have on the Earth's climate. They include sections on changes in the concentrations of atmospheric greenhouse gases; modelling of the global climate system; observed climate change over the last century; the detection of climate change due to human activities; changes in global sea levels due to global warming; the response of ecosystems to global climate change; and the research required to narrow the uncertainties in future predictions of global climate change.

Kemp, D.D. 1993. *Global Environmental Issues: A Climatological Approach*, second edition. London: Routledge.
This text examines the nature of global problems associated with climate change. It includes good sections on the greenhouse effect, acid deposition, ozone depletion, drought and possible effects of a nuclear winter. It is appropriate for students of environmental studies and geography.

Parry, M. 1990. *Climate Change and World Agriculture*. London: Earthscan.
This provides a good account of the likely patterns of change in climate and world agriculture as a consequence of global warming. It is a useful text for students, instructors and practitioners with interests in the effects of global warming.

Paterson, M. 1996. *Global Warming and Global Politics*. London: Routledge.
This book looks at the major theories within the discipline of international relations, and considers the emergence of global warming as a political issue.

O'Riordan, T. and Jager, J. (eds) 1995. *Politics of Climate Change*. London: Routledge.
This edited volume provides a critical analysis of the political, moral and legal responses to climate change in the midst of significant socio-economic policy shifts. In addition, it examines how climate change was put on the policy agenda of the EU, and how the United Nations Framework Convention and the subsequent Conference of Parties evolved.

Whyte, I. 1995. *Climatic Change and Human Society*. London: Edward Arnold.
This book examines the various ways that climatic change can interact with society. It is useful for students of geography, politics and/or economics.

Essay questions

1 Discuss the potential roles played by anthropogenic emissions of greenhouse gases in contributing to global warming.

2 Discuss the arguments against any evidence for global warming.

3 Describe a General Circulation Model (GCM) and discuss the potential limitations of such models in accurately predicting future global climate change.

4 Describe the effects of volcanic activity on influencing global climate change.

5 Do you believe that international agreements will successfully tackle the anthropogenic ozone-depleting gases? Illustrate your answer with reference to past international treaties and agreements.

6 Examine the links between urban pollution and increased virulence of asthma.

7 'The most serious threat to the welfare of people in the next century is posed by the rising emissions of greenhouse gases; unless these are significantly reduced in volume, regional, and possibly global, disasters will be unavoidable!' Discuss.

8 Evaluate the evidence for human-induced global warming.

9 Describe the dynamics of the El Niño Southern Oscillation and discuss the likely consequences should the frequency of El Niño events increase due to global warming.

10 Describe how global warming may trigger a cascade of hazard effects.

11 'It is unlikely that the developed and developing nations will meet the targets set by the various international agreements on atmospheric pollution.' Discuss.

12 Describe the carbonate–silicate geochemical cycle and discuss how this has helped to stabilise global climate over the last 3,800 million years.

Multiple-choice questions

Choose the best answer for each of the following questions.

1 Triatomic oxygen that forms and is present near ground level to heights of about 12–15 km in the atmosphere is called:
  (a) tropospheric ozone
  (b) stratospheric ozone
  (c) atmospheric ozone
  (d) ozonosphere

2 The greatest concentration of ozone in the atmosphere can be found at an altitude of:
  (a) 5 km
  (b) 15 km
  (c) 25 km
  (d) 35 km

3 Organic compounds containing chlorine and bromine are called:
  (a) hydrocarbons
  (b) halocarbons
  (c) chlorocarbons
  (d) halons

4 The maximum loads of a pollutant that the environment can sustain before damage occurs is known as the:
  (a) critical load
  (b) threshold level
  (c) saturation load
  (d) a, b and c

5 A negatively charged molecule comprising one atom of hydrogen and two of oxygen is known as a:
  (a) hydroperoxyl radical
  (b) hydroxyl radical
  (c) hydrogen radical
  (d) a, b and c

6 Dimethyl sulphide is a biologically produced organic compound that may act as:
  (a) a condensation nuclei
  (b) a greenhouse gas
  (c) a catalyst for ozone dissociation
  (d) a, b and c

7 The effect that a given amount of a trace gas can have on forcing climate compared with the effect by the same amount of CO_2 is known as:
  (a) greenhouse effect
  (b) global warming effect
  (c) greenhouse warming potential
  (d) global warming potential

8 Which of the following probably has made the major contribution to sea-level rise over the past 100 years?
  (a) thermal expansion of the oceans
  (b) melting glaciers and small ice caps
  (c) melting of the Greenland ice sheet
  (d) melting of the Antarctic ice sheet

9 Which of the following has the highest GWP over a period of 20 years?
  (a) CO_2
  (b) CH_4
  (c) CFC-12
  (d) HCFC-22

10 Which of the following countries has the highest net greenhouse gas emissions?
  (a) China
  (b) Brazil
  (c) Canada
  (d) USA

11 Which of the following satellites monitors O_3 levels?
  (a) Landsat
  (b) Spot
  (c) Nimbus 7
  (d) EOS

12 Volcanic eruptions can result in a global cooling of between:
  (a) 0.2 and 0.5 °C
  (b) 0.5 and 2 °C
  (c) 2 and 5 °C
  (d) 5 and 10 °C

13 If greenhouse emissions continue at the current rate, global warming will cause a sea-level rise of 50–60 cm by the year:
  (a) 2000
  (b) 2025
  (c) 2050
  (d) 2100

14 The IPCC suggests that a doubling of atmospheric CO_2 is likely to increase the global mean surface temperature by between:
  (a) 1.5 and 4.5 °C
  (b) 2.5 and 5.5 °C
  (c) 3.5 and 6.5 °C
  (d) 4.5 and 7.5 °C

15 During an El Niño event the level of atmospheric CO_2:
  (a) decreases
  (b) increases
  (c) stays the same
  (d) increases and decreases rapidly

16 Arctic geotherms provide evidence in the form of thermal anomalies to suggest a warming over the last century in the order of:
  (a) 0.1–0.5 °C
  (b) 0.5–1 °C
  (c) 1–2 °C
  (d) 2–4 °C

17 The Earth's climate has been stable as far back as:
  (a) 3,800 Ma
  (b) 600 Ma
  (c) 2.4 Ma
  (d) 850,000 years ago

18 Which of the following volcanic gases is most likely to form an ion that may act as a catalyst for ozone-depleting reactions?
  (a) CO_2
  (b) SO_2
  (c) NH_3
  (d) H_2O

19 Significant depletion of stratospheric ozone concentrations were first recorded over the Antarctic in:
  (a) 1970
  (b) 1977
  (c) 1983
  (d) 1987

20 A large increase in the concentrations of atmospheric CO_2 may result in:
  (a) a depletion of stratospheric O_3 concentrations
  (b) an increase in stratospheric O_3 concentrations
  (c) a stabilisation of stratospheric O_3 concentrations
  (d) no effect on stratospheric O_3 concentrations

21 Which of the following types of electromagnetic radiation has the shortest wavelength?
  (a) X-rays
  (b) ultraviolet radiation
  (c) infrared radiation
  (d) radio waves

22 The largest volcanic eruption during the twentieth century was:
- (a) Mount Pinatubo
- (b) Toba
- (c) Krakatau
- (d) El Chichón

23 Which of the following gases is poisonous?
- (a) carbon dioxide
- (b) tropospheric ozone
- (c) CFCs
- (d) all three

24 The Intergovernmental Panel on Climate Change was set up by the:
- (a) WMO and UNEP
- (b) UN
- (c) World Bank
- (d) UNESCO

25 In June 1992, at the United Nations Conference on the Environment and Development in Rio de Janeiro, many countries agreed to support the UN Framework Convention on Climate Change, committing them to reducing emissions of the greenhouse gases CO_2, CH_4 and N_2O by the year 2000 to their:
- (a) pre-industrial levels
- (b) 1970 levels
- (c) 1980 levels
- (d) 1990 levels

Figure questions

1 Figure 3.4 shows the changes since the middle of the eighteenth century in the atmospheric concentrations of CFC-11, methane, carbon dioxide and nitrous oxide. Answer the following questions.
- (a) Which of the curves is not likely to increase drastically over the next few decades? Give reasons for your answer.
- (b) Explain why there is no atmospheric concentration of the gas shown on the fourth graph (in the bottom right-hand corner) before 1950.
- (c) Why do the first three graphs show a sudden increase after 1950?

2 Figure 3.14 shows the variation of global mean combined land–air and sea-surface temperatures from 1861 through 1989, plotted relative to the average (0.0) for the years 1951 through 1980. Answer the following questions.
- (a) Describe the main characteristics of the pattern of change.
- (b) Do these data show how human activity has caused global warming? Explain your answer.
- (c) Is the global mean combined land–air and sea-surface temperature an appropriate way of assessing global warming?

3 Figure 3.15 shows the best estimate (solid line) of sea-level rise between 1990 and 2100, with a range of uncertainty (grey area) for a modified version of the IPCC business-as-usual emissions scenario. Answer the following questions.
- (a) List the main contributions to sea-level rise.
- (b) Why is there such a large uncertainty in the predictions?
- (c) What are the main environmental consequences of such a rise in sea level?

Short questions

1 What effect does the reduction of stratospheric ozone have on climate?

2 How are stratospheric ozone levels being affected by human activity?

3 What is the 'greenhouse effect'?

4 How do volcanic eruptions affect climate?

5 What are global warming potentials and residence times?

6 List the main international actions on global atmospheric pollution that have resulted in attempts to reduce ozone depletion and greenhouse gas emissions.

7 Describe the role that clouds may play in the climate system.

8 There has been between a 10 and 20 cm rise in sea level over the last 100 years. What are the contributions to this sea-level rise, and estimate how much is attributed to each set of contributing factors?

9 What is a GCM?

10 Describe the likely consequences if human emissions result in the doubling of the atmospheric concentrations of carbon dioxide.

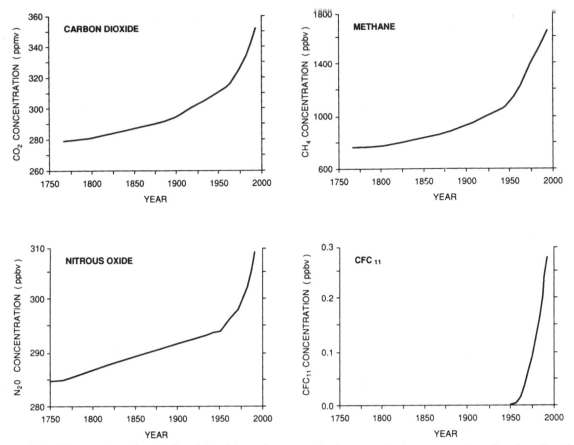

Figure 3.4 *Changes since the middle of the eighteenth century in the atmospheric concentration of carbon dioxide, methane, nitrous oxide and the commonly occurring CFC, CFC-11. Over the last few decades there has been a very large increase in the atmospheric concentrations of CFCs, which were absent before the 1930s. After IPCC (1990).*

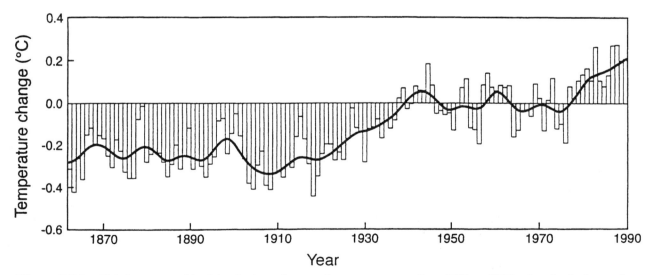

Figure 3.14 *Global mean combined land–air and sea-surface temperatures from 1861 to 1989, plotted relative to the average (0.0) for the years 1951 to 1980. Note that the rise in temperature has not taken place at a consistent rate: noticeable increases occurred between 1910 and 1940, and since the early 1970s (after Houghton et al. 1995).*

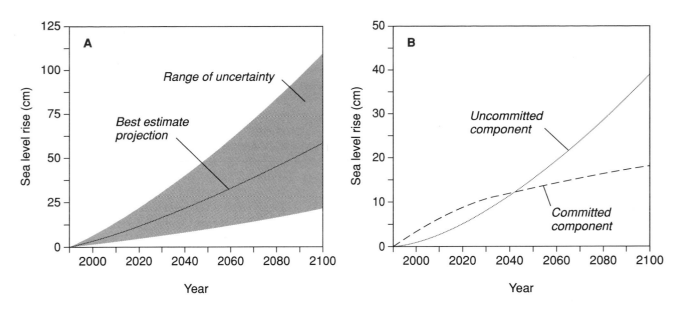

Figure 3.15 *1994 IPCC report predictions for the amount by which global sea level will rise between the years 1990 and 2100 under a mid-range rate of increased greenhouse gas emissions, the business-as-usual scenario (BAU), which produces the best estimate (solid line) of a 58 cm rise in global (eustatic) sea level by 2100, with a range of uncertainty (grey area) from a 21 to a 105 cm rise by 2100. Redrawn after Houghton et al. (1995).*

Answers to multiple-choice questions

1 a; 2 c; 3 b; 4 a; 5 a; 6 a; 7 d; 8 b; 9 c; 10 d; 11 c; 12 a; 13 d; 14 a; 15 b; 16 d; 17 a; 18 b; 19 b; 20 a; 21 a; 22 a; 23 b; 24 a; 25 d.

Answers to figure questions

1 (a) The curve on the fourth graph (in the bottom right-hand corner) is not likely to increase drastically during the next few decades because of the ban and restricted use of CFCs. (b) No emissions are shown on the fourth graph because this graph represents the variation in CFC concentrations over the last few centuries, and since CFCs were not produced prior to the 1950s there are no emissions. (c) The first three graphs show a sudden increase in emissions after the 1950s because of increased emissions associated with increased industrialisation in the post-war years.

2 (a) The main pattern of change is best seen using the curve showing the running mean of temperature. The curve shows that prior to the 1920s temperature was relatively stable and then rose sharply, by about 0.3 °C, until the late 1940s. Temperature stabilised to the early 1980s and then rose sharply again. In summary, there appears to be a net warming during the twentieth century by about 0.5 °C. (c) This is probably the most appropriate measure of

global temperature change, because it considers the whole surface of the Earth, particularly the major store of heat energy, that is, the oceans. It considers only the surface of the oceans, however, and much heat may be stored/lost from the deep seas. In addition, averaging the overall temperature change may not be appropriate since the changes in contrast in temperature between areas may be more important in changing climatic patterns than the average global temperature.

3 (a) The main contributions to sea-level rise are thermal expansion of the oceans; melting ice sheets (Greenland and Antarctic); and melting glaciers and ice caps. (b) The large uncertainty in predicting sea-level change is due to the uncertainty associated with calculating the amount of global warming that may occur in the next century and due to the complex nature of feedback systems that may offset the effect of the increased temperatures on the melting of glaciers in polar regions. (c) The main environmental consequences of sea-level rise include flooding of coastal regions; salt-water intrusion into ground-water sources; and changing coastal sedimentation patterns.

Answers to short questions

1 Reducing stratospheric O_3 levels results in a cooling of the stratosphere and a warming of the

troposphere. The net result is a cooling of the Earth's surface, but the balance is strongly influenced by latitude, altitude and seasons.

2 Stratospheric ozone levels are being reduced by anthropogenic emissions of CFCs and other chemicals such as nitrous oxide. Significant depletion in stratospheric ozone concentrations have been recorded since 1977. Depletions are now common over the poles following the springs for each respective pole.

3 The 'greenhouse effect' warms the Earth's atmosphere. This involves short-wavelength (ultraviolet) solar radiation reaching the Earth's surface and being re-radiated as long-wavelength (infrared) radiation back up into the atmosphere, where greenhouse gases absorb the radiation and warm the Earth.

4 Volcanic eruptions emit huge quantities of aerosols into the atmosphere, which reduces incoming solar radiation and leads to global cooling, usually of between 0.2 °C and 0.5 °C, for several years. Volcanic eruptions also emit greenhouse gases, for example carbon dioxide, and ozone-depleting gases (e.g. Cl_2, F_2). In addition, volcanic activity emits gases that may act as catalysts for ozone-depleting reactions (SO_2, which forms H_2SO_4), and cause either negative or positive feedback mechanisms for global climate change.

5 A global warming potential is the effect that a given amount of a trace gas can have on forcing climate compared with the effect of the same amount of CO_2. Residence time is the period that a particular gas will remain active in the atmosphere.

6 The main international actions on global atmospheric pollution that have resulted in attempts to reduce ozone depletion include the Montreal Protocol, 1987, with amendments in Helsinki, 1989, and Copenhagen 1992; and for reductions in greenhouse gases the UN Framework Convention on Climate Change, 1992.

7 Clouds are probably the most important self-regulating mechanism in controlling feedback in the ocean–atmosphere system, and preventing a runaway greenhouse or icehouse effect. Their warming effect may be many times greater than doubling the CO_2 levels of a region. Also, their overall cooling effect makes the Earth 10–15 °C cooler than it would otherwise be if the planet were cloudless. There are two major sets of clouds: tropical clouds, which

reflect sunlight back into the atmosphere to cool the system and which also have a greenhouse warming effect; and middle- to high-latitude clouds, which have a net cooling effect. Clouds also provide extra hydroxyl molecules, which are capable of oxidising CH_4 and NO_x, thereby removing some of the greenhouse gases from the atmosphere and reducing their greenhouse effect.

8 The IPCC (1990) estimates that about 2 to 4 cm of rise has been due to thermal expansion of the oceans, 1.5 to 7 cm to the melting of small glaciers and ice caps, 1 to 4 cm to melting of the Greenland Ice Sheet, and −5 to +5 cm to melting of the Antarctic Ice Sheet.

9 A GCM is an abbreviation for General Circulation Model. It is a simulation of atmospheric circulation involving a system of equations used to describe atmospheric circulation and ocean-water motion, the heat exchange and fluxes within this system, and the consequences. GCMs usually involve the solution of these equations on a high-speed computer or super-computer.

10 A doubling of atmospheric carbon dioxide concentrations is likely to increase the global mean surface temperature by between 1.5 and 4.5 °C and result in increased storminess. This increase in temperature is likely to result in global sea-level rise. It is mainly due to the thermal expansion of the oceans and melting of mountain glaciers and ice caps, probably in the order of 50 to 60 cm. Estimating the effects is difficult, however, because of the complex feedbacks involved in the climate system that are not fully understood.

Additional references

Behrenfield, M.J., Bale, A.J., Kolber, Z.S., Aiken, J. and Falkowski, P.G. 1996. Confirmation of iron limitation of phytoplankton photosynthesis in the equatorial Pacific Ocean. *Nature*, 383, 508–11.

Coale, K.H., Johnson, K.S., Fitzwater, S.E., Gordon, R.M., Tanner, S., Chavez, F.P., Ferioli, L., Sakamoto, C., Rogers, P., Millero, F., Steinberg, P., Nightingale, P., Cooper, D., Cochlan, W.P., Landry, M.R., Constantinou, J., Rollwagen, G., Trasvina, A. and Kudela, R. 1996. A massive phytoplankton bloom induced by an ecosystem-scale iron fertilization experiment in the equatorial Pacific Ocean. *Nature*, 383, 495–501.

Cooper, D.J., Watson, A.J. and Nightingale, P.D. 1996. Large decrease in ocean-surface CO_2 fugacity in response to *in situ* iron fertilization. *Nature*, 383, 511–13.

de Baar, H.J.W., de Jong, J.T.M., Bakker, D.C.E., Loscher, B.M., Veth, C., Bathmann, U. and Smetacek, V. 1995. Importance of iron for plankton blooms and carbon dioxide drawdown in the Southern Ocean. *Nature*, 373, 412–15.

Edouard, S., Legras, B., Lefèvre, F. and Eymard, R. 1996. The effect of small-scale inhomogeneities on ozone depletion in the Arctic. *Nature*, 384, 444–7.

Gribbin, J. and Gribbin, M. 1996. The Greenhouse Effect. *New Scientist*, 6 July, Inside Science, 92, 4 pp.

Hammitt, J.K., Jain, A.K., Adams, J.L. and Wuebbles, D.J. 1996. A welfare-based index for assessing environmental effects of greenhouse-gas emissions. *Nature*, 381, 301–3.

Keeling, C.D., Chin, J.F.S. and Whorf, T.P. 1996. Increased activity of northern vegetation inferred from atmospheric CO_2 measurements. *Nature*, 382, 146–9. This paper explores the seasonal changes in atmospheric CO_2 levels in the Northern Hemisphere and attempts to disassemble the natural from the anthropogenic variability. In the Northern Hemisphere, the concentration of atmospheric CO_2 shows an increase during the winter months and a decline during the summer, mainly reflecting the seasonal growth in land vegetation. In the far north the amplitude of this cycle varies between 15 and 20 ppmbv but decreases southward to approximately 3 ppmbv near the Equator. In a study of the atmospheric CO_2 record since the early 1960s, Keeling *et al.* show that the amplitude of the seasonal CO_2 cycle has increased by 20 per cent in Hawaii and by 40 per cent in the Arctic, accompanied by a lengthening of the growing season by about 7 days. The authors propose that the increased amplitude observed in the atmospheric CO_2 cycle is a consequence of the increased sequestration of CO_2 by land plants brought about by climate changes accompanying the recent temperature increases.

Santer, B.D., Taylor, K.E., Wigley, T.M.L., Johns, T.C., Jones, P.D., Karoly, D.J., Mitchell, J.F.B., Oort, A.H., Penner, J.E., Ramaswamy, V., Schwartzkopf, M.D., Stouffler, R.J. and Tett, S. 1996. A search for human influences on the thermal structure of the atmosphere. *Nature*, 382, 39–46.

A very useful overview of the marriage between observed spatial patterns of temperature change in the free atmosphere between 1963 and 1987 and the development of computer-based climate models. Vertical changes in the thermal structure of the atmosphere have been predicted from various computer models (Atmospheric General Circulation Models, AGCMs). The results reported in this paper suggest that the observed temperature changes between 1963 and 1987 are unlikely to be the result of natural internally generated variability of the climate system, but more likely anthropogenic influences (greenhouse gas emissions, direct sulphate aerosols and stratospheric ozone). Uncertainties remain, however, in the observational data, with a need for improved histories of radiative forcing due to natural and anthropogenic factors coupled to more robust numerical modelling of anthropogenic climate-change signals versus natural factors.

Sokolik, I.N. and Toon, O.B. 1996. Direct radiative forcing by anthropogenic airborne mineral aerosols. *Nature*, 381, 681–3.

Turner, S.M., Nightingale, P.D., Spokes, L.J., Liddicoat, M.I. and Liss, P.S. 1996. Increased dimethyl sulphide concentrations in sea water from *in situ* iron enrichment. *Nature*, 383, 511–13.

In ecosystem-scale studies, Behrenfield *et al.*, Coale *et al.*, Cooper *et al.* and Turner *et al.* show that seeding with low concentrations of dissolved iron in surface waters in the equatorial Pacific Ocean led to a massive phytoplankton bloom, which sequestered large amounts of CO_2 and nitrates. Thus, this study supports the previously published proposals that iron bioavailability is a rate-limiting step in biomass production in oceanic surface waters.

Table 3.1 *Summary of assumptions in the six IPCC 1992 alternative IS92 Scenarios.*

Scenario	Population	Economic growth	Energy supplies	Other	CFCs
IS92a	World Bank (1991) 11.3 billion by 2100	1990–2025: 2.9% 1990–2100: 2.3%	12,000 EJ conventional oil 13,000 EJ natural gas. Solar costs fall to $0.075/kWh. 191 EJ/year of biofuels available at $70/barrel†	Legally enacted and internationall agreed controls on SO_x, NO_x and NMVOC emissions. Efforts to reduce emissions of SO_x, NO_x and CO in developing countries by middle of next century.	Partial compliance with Montreal Protocol. Technological transfer results in gradual phase out of CFCs in non-signatory countries by 2075.
IS92b	World Bank (1991) 11.3 billion by 2100	1990–2025: 2.9% 1990–2100: 2.3%	Same as 'a'	Same as 'a' plus commitments by many OECD countries to stabilise or reduce CO_2 emissions.	Global compliance with scheduled phase out of Montreal Protocol.
IS92c	UN Medium-Low Case 6.4 billion by 2100	1990–2025: 2.0% 1990–2100: 1.2%	8,000 EJ conventional oil 7,300 EJ natural gas Nuclear costs decline by 0.4% annually	Same as 'a'	Same as 'a'
IS92d	UN Medium-Low Case 6.4 billion by 2100	1990–2025: 2.7% 1990–2100: 2.0%	Oil and gas same as 'c' Solar costs fall to $0.065/kWh 272 EJ/year of biofuels available at $50/barrel	Emission controls extended worldwide for CO, NO_x, NMVOC and SO_x. Halt deforestation. Capture and use of emissions from coal mining and gas production and use.	CFC production phase out by 1997 for industrialised countries. Phase out of HCFCs.
IS92e	World Bank (1991) 11.3 billion by 2100	1990–2025: 3.5% 1990–2100: 3.0%	18,400 EJ conventional oil Gas same as 'a' Phase out nuclear by 2075	Emission controls which increase fossil energy costs by 30%.	Same as 'd'
IS92f	UN Medium-High Case 17.6 billion by 2100	1990–2025: 2.9% 1990–2100: 2.3%	Oil and gas same as 'e' Solar costs fall to $0.083/kWh Nuclear costs increase to $0.09/kWh	Same as 'a'	Same as 'a'

† Approximate conversion factor: 1 barrel = 6 GJ.

Source: After Houghton *et al.* (1995)

Table 3.2 *Selected results of the six IS92 Scenarios.*

Scenario	Scenario year	Emissions per year*			
		CO_2 (GtC)	CH_4 (Tg)	N_2O (TgN)	S (TgS)
IS92a	1990	7.4	506	12.9	98
	2025	12.2	659	15.8	141
	2100	20.3	917	17.0	169
IS92b	2025	11.8	659	15.7	140
	2100	19.1	917	16.9	164
IS92c	2025	8.8	589	15.0	115
	2100	4.6	546	13.7	77
IS92d	2025	9.3	584	15.1	104
	2100	10.3	567	14.5	87
IS92e	2025	15.1	692	16.3	163
	2100	35.8	1072	19.1	254
IS92f	2025	14.4	697	16.2	151
	2100	26.6	1168	19.0	204

* The figures for CO_2 are anthropogenic emissions. The figures for CH_4, N_2O and S are combined natural and anthropogenic emissions. Natural emissions in 1990 are estimated as: CH_4 = 340 $TgCH_4$, N_2O = 4.7 TgN, and S = 74 TgS. It is generally assumed that natural emissions will remain constant; thus anthropogenic CH_4, N_2O and S emissions for any scenario and year can be estimated by subtracting the figures for natural emissions presented in the text from the numbers in the body of the table.

Source: After Houghton *et al.* (1995)

Aims

- To examine the formation and deposition of acid precipitation.

- To assess the environmental effects and distribution of acid precipitation.

- To evaluate the methods and means of reducing the effects of acid precipitation and the remedial measures that may be undertaken to mitigate the effects.

Key point summary

- Human activities result in emissions of various pollutants (principally SO_2, NO_x, NH_3, hydrocarbons and particulate matter) into the atmosphere. This results in environmental problems such as poor air quality and **acidic deposition**.

- Acidic deposition as a result of natural and human activities is formed by the reaction of CO_2 with H_2O (**carbonic acid, H_2CO_3**), SO_2 with H_2O (**sulphuric acid, H_2SO_4**) and NO_x with H_2O (**nitric acid, HNO_3**). The **pH** of some human-induced acidic deposition is as low as 2.

- Acidic deposition is a particular problem in industrialised regions and countries where the combustion of fuels releases large quantities of SO_2 into the atmosphere. It is also a problem in countries down-wind from the polluter.

- Acidic deposition causes acidification of ground waters and surface waters, damage to life, particularly forests and aquatic life, and building decay. Acid ground water may cause corrosion of pipes to mobilise toxic metals such as Pb, Cu, Cd and Al.

- Buffering reactions due to the presence of certain clay minerals, and because of cations in water in some soils and lakes, may reduce the immediate effects of acidic deposition on those environments.

The susceptibility of soils and lakes to acidification is quantified as the **acid susceptibility** and **acid-neutralising capacity**, respectively.

- Recovery from acidification depends on the sensitivity of the ecosystem. Reducing acidification involves reducing anthropogenic emissions, mainly from fossil-fuel-burning power stations, e.g. using appropriate clean technologies such as atmospheric fluidised-bed combustion, the use of active coke and clean combustion engines in both domestic and commercial vehicles. Incentives and legislation are needed to implement such technologies.

- International conventions and agreements have been signed in order to reduce poor air quality and the emissions of SO_2. These include the Convention on Long-Range Transboundary Air Pollution in the early 1970s; Conference on the Acidification of the Environment, Stockholm 1982; the 30% Club, introduced in Ottawa in 1984; the Multilateral Conference on the Environment, in Munich in 1984; the Helsinki Meeting in 1985; the International Conference on Acidification and its Policy Implications, held in Amsterdam in 1986; and the Sofia Meeting of 1988.

Main learning hurdles

Ions

Students with a poor science background may not fully appreciate what is meant by ion, cation and anion. The instructor should describe the structure of an atom and explain how it becomes positively or negatively charged.

pH

This is commonly a difficult concept for students who lack a strong science background to understand fully because they have been confused with definitions such as pH = –log [H$^+$]. The instructor should explain such definitions, but emphasise that much of the discussion within this chapter simply requires a basic appreciation that low values of pH mean that a solution is acidic whereas high values are alkaline.

Chemical buffering

This is a difficult concept for students with a poor chemistry background to understand. The instructor should explain the meaning of buffering and why it is important in understanding acid susceptibility.

Key Terms

Acid ground; acid susceptibility; acid-neutralising capacity (ANC); acidic deposition; atmospheric fluidised-bed combustion; buffering reaction; dry deposition; forest death; hydroxyl radical; pH; photochemical smog; photon; wet deposition.

Issues for group discussion

Discuss the effectiveness of policies to reduce the effects of acid deposition.
The students should review the international conventions on the reduction of air pollution and they should read Brydges and Wilson (1991). Students should emphasise the effectiveness of legislation, and the need for governments to implement incentives to encourage the use of clean technologies, as well as proper provisions for prosecuting offenders.

Assess the effect of acidic deposition with reference to a specific ecosystem.
The students should find information on a specific ecosystem, for example a boreal forest, a high-latitude lake, or tundra ecosystem. The students should discuss the nature of vegetation, soil, surface and ground waters in these environments and describe how acidic deposition can affect each of these components, as well as how they interrelate.

Describe the problems associated with atmospheric pollution in urban environments and discuss possible means of reducing the problem.
The discussion should examine the effects of all types of urban pollution, including acidic deposition, smog and tropospheric ozone, on the quality of life, environmental health, urban ecosystems and buildings.

The main methods of pollution reduction may involve clean energy production and clean industrial activity, energy conservation, and reduced vehicle emission. Students should emphasise how governments implement these methods.

Selected readings

Battarbee, R. 1992. Holocene lake sediments, surface water acidification and air pollution. *Quaternary Proceedings*, 2, 101–10.
This paper shows an example of the use of proxy evidence to examine the timing and nature of lake acidification. It also illustrates how such studies are important for the establishment of baseline conditions for environmental management.

Brydges, T.G. and Wilson, R.B. 1991. Acid rain since 1985 – times are changing. In: Last, F.T. and Watling, R. (eds), *Acid Deposition: Its Nature and Impacts*. Edinburgh: The Royal Society of Edinburgh, 1–16.
This paper provides a review of the multinational environmental issues surrounding acidic deposition. It describes the development of scientific knowledge about acid deposition and the public awareness of the problem, as well as the international action concerning the pollution control programmes.

Pearce, F. 1993. How Britain hides its acid soil. *New Scientist*, 27 February, 29–33.

Reuss, J.O., Cosby, B.J. and Wright, R.F. 1987. Chemical processes governing soil and water acidification. *Nature*, 329, 27–32.
This is an important paper because it was one of the first that really began to examine and understand the complex chemical processes involved in soil and water acidification.

Textbooks

Carter, F.W. and Turnock, D. (eds) 1996. *Environmental Problems in Eastern Europe*. London: Routledge.
This edited volume analyses the major forms of pollution and the resulting decline in the quality of life in each country in Eastern Europe. Air, water, soil and vegetation pollution, dumping of waste, nuclear power and transboundary issues are considered in relation to the role of legislation, political movements, international co-operation, aid and education that strives for solutions to environmental problems. This

is a particularly interesting read for students who are studying geography and/or environmental politics.

Journal of the Geological Society of London, 1986. *Geochemical Aspects of Acid Rain*. Thematic set of papers, 619–720.
This is a useful thematic set of research papers on acid deposition. It is somewhat specialised, but it is recommended for an in-depth appreciation of the topic.

Last, F.T. and Watling, R. (eds) 1991. *Acid Deposition: Its Nature and Impacts*. Edinburgh: The Royal Society of Edinburgh.
This is a useful volume on acid deposition, produced as an edited conference proceedings. It is an essential read for environmental science students, instructors and researchers with interests in acidic deposition.

Kamari, J., Brakke, D.F., Jenkins, A., Norton, S.A. and Wright, R.F. (eds) 1989. *Regional Acidification Models: Geographical Extent and Time Development*. London: Springer-Verlag.
This edited text reviews the development and use of mathematical models for the assessment of regional acidification and other effects of air pollutants on the environment. It comprises aspects of sensitivity distribution, time evolution of regional impacts, and uncertainty in model applications.

McCormack, J. 1989. *Acid Earth: The Global Threat of Acid Pollution*. London: Earthscan.
This is an easy read that will appeal to students who wish to understand the nature of acid deposition and the international context in which nations have sought to reduce the associated problems.

Pearce, F. 1987. *Acid Rain*. Harmondsworth/New York: Penguin Books.
This readable text examines acid deposition. It discusses the effects on the natural environment, people's health, the corrosion of building materials, acidification of water sources, and policy issues relating to national and international efforts aimed at reducing the emissions that cause acidic deposition.

Essay questions

1 What is acidic deposition, and what are its consequences for the environment?

2 Describe the factors that determine the susceptibility of a region to the effects of acid deposition.

3 Discuss the meaning of the terms acid-neutralising capacity (ANC) and acid susceptibility.

4 Discuss the reversibility of the regional effects of acidic deposition.

5 Discuss the various international conventions and agreements aimed at reducing atmospheric pollution that leads to poor air quality and acidic deposition.

6 Evaluate the effectiveness of the various methods used to reduce the effects of acidic deposition in cities.

7 Describe the ecological changes that occur in a temperate forest as a result of acidic deposition.

8 Describe the proxy data that can be used to examine the history of acid deposition in mid- and high-latitudes.

9 Explain the distribution of acid deposition and its effects in EITHER Europe OR North America.

10 Describe the environmental health hazards from acidic deposition.

11 Evaluate the usefulness of diatoms in assessing the anthropogenic acidification of lakes.

12 Assess the effectiveness of the various mechanisms for helping to reduce surface-water acidification.

Multiple-choice questions

Choose the best answer for each of the following questions.

1 The ability of a water body to reduce the acidity of incoming acid water is referred to as:
  (a) biochemical oxygen demand
  (b) buffering capacity
  (c) cation exchange capacity
  (d) acid-neutralising capacity

2 A chemical substance that releases hydrogen ions when dissolved in water is known as an:
  (a) acid
  (b) alkali
  (c) anoxic
  (d) anaerobic

3 Which of the following formulae is carbonic acid?
  (a) H_2CO_3
  (b) HCO_3
  (c) HCl
  (d) H_2Cl_3

4 A chemical that can maintain the pH of a solution by reacting with excess acid or alkali is known as a:
  (a) neutraliser
  (b) catalyst
  (c) buffer
  (d) dampener

5 The incorporation of particles and gases into rain and snow which deposit by gravity is known as:
  (a) wet deposition
  (b) dry deposition
  (c) suspension deposition
  (d) particle deposition

6 Soft water is water that:
  (a) lacks or has very low concentrations of dissolved salts such as calcium, magnesium and other ions in solution
  (b) contains high concentrations of dissolved salts such as calcium, magnesium and other ions in solution
  (c) contains radon
  (d) contains organic pollutants

7 Which of the following contributes the largest amounts of nitrogen oxides to the atmosphere?
  (a) tropospheric aircraft
  (b) fossil-fuel combustion
  (c) lightning
  (d) soils

8 The extreme smog over Athens on 1 October 1991 was known as:
  (a) *nephos*
  (b) pea souper
  (c) skeggy
  (d) depressor

9 Which of the following gases is not emitted by volcanoes?
  (a) NO_x
  (b) F_2
  (c) CFCs
  (d) CO_2

10 At the Sofia Meeting in 1988 it was agreed to reduce NO_x emissions by 1998 by:
  (a) 10 per cent
  (b) 30 per cent
  (c) 50 per cent
  (d) 70 per cent

11 Which of the following metals is mobilised by the corrosion of water pipes?
  (a) aluminium
  (b) copper
  (c) cadmium
  (d) a, b and c

12 On 14 June 1994, 29 countries signed a treaty in Oslo to reduce acidic deposition pollution by:
  (a) 2000 to 30 per cent of the 1980 levels
  (b) 2000 to 80 per cent of the 1980 levels
  (c) 2010 to 30 per cent of the 1980 levels
  (d) 2010 to 80 per cent of the 1980 levels

13 In the USA the main legislation against air pollution is the:
  (a) 1963 Clean Air Act
  (b) Convention on Long-Range Transboundary Air Pollution
  (c) Air Pollution Act
  (d) a, b and c

14 One of the first to recognise the effects of acidic deposition was:
  (a) Robert Smith
  (b) Richard Battarbee
  (c) Charles Darwin
  (d) Lewis Owen

15 The infamous smog of December 1952 in London resulted in the death of about:
  (a) 10 people
  (b) 100 people
  (c) 1,000 people
  (d) 4,000 people

16 Acidic deposition has pH values that can be as low as:
  (a) 1
  (b) 2
  (c) 7
  (d) 12

17 What percentage of sulphur deposition in Norway during the years 1978–1982 was from external sources?
  (a) <10 per cent
  (b) 25 per cent
  (c) 75 per cent
  (d) >90 per cent

18 The bulbous precipitates that form beneath anvils and gargoyles as a result of the reaction of acidic deposition and limestone comprise:
  (a) $CaCO_3$
  (b) $CaMgCO_3$
  (c) $CaCl_2$
  (d) $CaSO_4$

19 Secondary pollutants are produced:
  (a) directly from industrial and domestic activities
  (b) in the atmosphere by chemical processes acting on primary pollutants
  (c) by the dissociation of tertiary pollutants by chemical processes acting on primary pollutants
  (d) as they are deposited in water bodies by chemical processes acting on primary pollutants

20 Which of the following is true?
  (a) wet deposition affects a larger area than dry deposition
  (b) dry deposition affects a larger area than wet deposition
  (c) wet and dry deposition affect similar size areas
  (d) none of the above statements

21 Acidic deposition affects trees in the following way:
  (a) the tree does not get enough important nutrients since they have been leached from the ground
  (b) acid deposition affects the leaves and damages cells
  (c) by affecting the symbiotic fungi that colonise the roots of trees, providing both nutrients and helping to protect the trees against disease
  (d) a, b and c

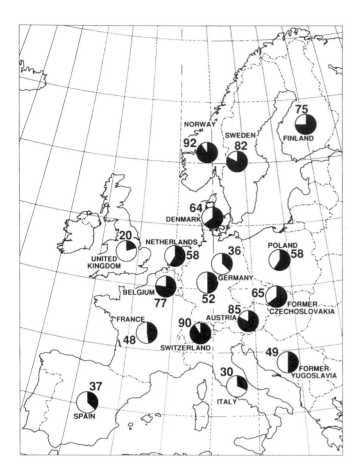

Figure 4.3 *The contribution, shown as a percentage, of external sources (solid portions in pie diagrams) to the amount of sulphur deposited in some European countries. Percentages are based on EMEP model calculations for 1978–1982. Redrawn after the National Environment Protection Board (1987).*

22 Which of the following acronyms refers to the best economical method of clean-up:
- (a) BMA
- (b) BAT
- (c) BTALEC
- (d) BATNEEC

23 Dimethyl sulphide is produced by:
- (a) chemical processes in the atmosphere acting on primary pollutants
- (b) chemical reactions during the production of energy at coal-fired power stations
- (c) marine plankton
- (d) bacteria in acidified soils

24 Which of the following statements is false?
- (a) Acidic deposition could act as a negative feedback mechanism to reduce greenhouse warming
- (b) Acidic deposition is solely the result of human activity
- (c) More than 50 per cent of the world-wide global sulphur- and nitrogen-related pollution comes from Europe
- (d) All of the above

25 Which of the following microfossils are the best indicators of the pH of aquatic environments?
- (a) pollen
- (b) foraminifera
- (c) diatoms
- (d) coccoliths

Figure questions

1 Figure 4.3 shows the contribution, shown as a percentage, of external sources (solid portions) to the amounts of sulphur deposited in some European countries. Answer the following questions.
- (a) Why are the percentages of external sources of sulphur deposits in Norway, Sweden, Switzerland and Austria so much higher than other countries?
- (b) What are the environmental effects of sulphur deposition?
- (c) Suggest ways to reduce the absolute amounts of sulphur deposition.

2 Figure 4.10 shows the changes in diatom assemblages and reconstructed pH for a sediment core from the Round Loch of Glenhead, Scotland. Answer the following questions.

- (a) Describe the characteristics of changes in the diatom fauna of the lake over the last 150 years.
- (b) What do these data suggest?
- (c) Why are these data environmentally important?

Short questions

1 List the major gases and the acids that they produce that are associated with acidic deposition.

2 Describe the nature and characteristics of photochemical smog.

3 What is pH and how is it measured?

4 List the main conventions and agreements to help to reduce poor air quality and the emissions of SO_2.

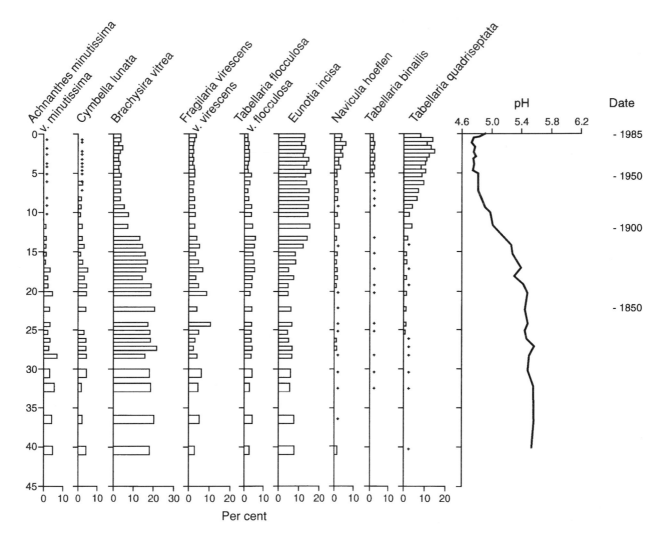

Figure 4.10 *Changes in diatom assemblages and reconstructed pH for a sediment core from the Round Loch of Glenhead, Scotland. Dates are interpolated from ^{210}Pb ages. Redrawn after Jones et al. (1989) and Birks et al. (1990) in Battarbee (1983).*

5 What is soft water?

6 Describe the effects acidic deposition has on trees.

7 What is a buffer?

8 Describe the extent of the different types of acidic deposition.

9 What is ANC?

10 Suggest means of reducing acidic deposition.

Answers to multiple-choice questions

1 d; 2 a; 3 a; 4 c; 5 a; 6 a; 7 b; 8 a; 9 c; 10 b; 11 d; 12 d; 13 a; 14 a; 15 d; 16 b; 17 d; 18 d; 19 b; 20 a; 21 d; 22 d; 23 c; 24 b; 25 c.

Answers to figure questions

1 (a) The percentages of external sources of sulphur deposits in Norway, Sweden, Switzerland and Austria are higher than other countries because the pollutants from the adjacent heavily industrialised regions are carried by air systems into these areas. In the case of Norway and Sweden, westerly weather systems blow pollutants from Britain and Germany, and in the case of Switzerland and Austria pollutants are forced northwards from northern Italy. (b) The environmental effects of sulphur deposition include acidification of surface and ground waters, and soils; destruction to vegetation and aquatic life; smogs such as Arctic haze; and building damage. (c) The absolute amounts of sulphur deposition can be reduced by cutting down anthropogenic emissions, mainly from

fossil-fuel-burning power stations, for example by using appropriate clean technologies such as atmospheric fluidised-bed combustion, and active coke, and cleaning combustion engines in both domestic and commercial vehicles.

2 (a) Prior to the late nineteenth century there is little change in the diatom faunas; these change rapidly within a few decades to become dominated by two species, *Eunotia incisa* and *Tabellaria quadriseptata*. (b) Diatoms are very sensitive to pH, and these data suggest that the diatom fauna changes may be due to acidification of the lake. (c) These data are important because they prove that the lakes were not initially as acidic as today and that they became acidified during the early twentieth century.

Answers to short questions

1 The main gases that produce the acids that are associated with acidic deposition are carbon dioxide (CO_2) combining with water to form carbonic acid (H_2CO_3); sulphur dioxide (SO_2) combining with water to form sulphuric acid (H_2SO_4); nitrogen oxides (NO_x) combining with water to form nitric acid (HNO_3); and chlorine (Cl_2) combining with water to form hydrochloric acid (HCl).

2 Photochemical smog results in poor air quality and is caused by sunlight energy catalysing chemical reactions, mainly with nitrogen compounds and hydrocarbons, commonly to produce a reddish yellow-brown haze. Photochemical smog commonly forms on very warm and sunny days in large urban areas subject to large amounts of tailpipe (exhaust) emissions from motor vehicles.

3 pH is a logarithmic scale, ranging from 1 to 14, which provides a measure of the acidity or alkalinity of a solution. It is essentially the measure of the concentration of hydrogen ions [H^+] in a solution. A pH value of 7 is neutral, while decreasing values indicate increasing acidity, and values greater than 7 represent increasing alkalinity (basicity).

4 The main conventions and agreements that have been signed to help reduce poor air quality and emissions of SO_2 include the Convention on Long-Range Transboundary Air Pollution in the early 1970s; the Conference on the Acidification of the Environment in Stockholm, 1982; the 30% Club, introduced in Ottawa in 1984; the Multilateral Conference on the Environment, in Munich, 1984; the Helsinki Meeting, 1985; the International Conference on Acidification and its Policy Implications, held in Amsterdam in 1986; and the Sofia Meeting of 1988.

5 Soft water is water that lacks or has a very low concentration of dissolved salts such as calcium, magnesium and other ions in solution (commonly carbonates), which precipitate out to cause the 'furring up' of pipes and appliances.

6 Acidic deposition affects trees by reducing the available amounts of important nutrients that they need through leaching from the ground; causing cell damage due to acidic waters; and by affecting the symbiotic fungi that colonise the roots of trees, which provide both nutrients and help to protect the trees against disease.

7 A buffer is a chemical that can maintain the pH of a solution by reacting with the excess acid or alkali (base). Limestone, for example, is a natural buffer that helps to stabilise the pH of ground water and soil close to neutral.

8 Acidic deposition may be widespread. Dry deposition usually concentrates within several tens of kilometres from the polluter. Wet deposition, however, occurs over many tens to thousands of kilometres from the polluter.

9 ANC is an abbreviation for acid-neutralising capacity. It is the ability of a water body to reduce (or neutralise) the acidity of incoming acid water.

10 Reducing acidification involves cutting down anthropogenic emissions, mainly from fossil-fuel-burning power stations, for example by using appropriate clean technologies such as atmospheric fluidised-bed combustion, and active coke, and cleaning combustion engines in both domestic and commercial vehicles.

Additional reference

Critical Loads Advisory Group 1995. *Critical Loads of Acid Deposition for United Kingdom Freshwaters*. Prepared at the request of the Department of the Environment, 80 pp. ISBN: 1–870393–25–2.

Aims

- To emphasise the importance of water as a resource essential for life.

- To examine the nature and effects of the main waterborne environmental pollutants.

- To examine the various mechanisms by which pollutants can be introduced and transferred through the hydrological and associated systems.

- To assess the various mechanisms by which water pollution can be avoided, mitigated or remedied.

- To consider the political, social and economic implications and importance of clean water resources.

Key point summary

- Clean and abundant water is an important resource essential for life.

- Water shortages in many countries and human activities have polluted water resources throughout history. Many of the world's beaches are suffering severe pollution, some with dangerously high levels of toxic chemicals. There are many types of pollutants, both natural and created by human activities.

- An understanding of the **hydrological cycle** is important in comprehending the routes that pollutants can travel through the ecosphere, particularly through **food chains** and **food webs**.

- **Sewage** and **sewage sludge disposal** are one of the most common pollutants, generally being disposed of in an untreated form, or raw state, into streams and seas.

- Treatment of sewage involves filtering, biological digestion and drying to produce sewage sludge, which is easier to handle and dispose of. Sewage treatment also aims to destroy disease-carrying organisms, for example cholera, typhoid, cryptosporidia, leptospirosis and giardia. Sewage sludge properly treated can provide a valuable fertiliser, or energy source through incineration to generate electricity. The decay of organic matter produces NH_3 and nitrates, which, if in sufficient concentrations, may contaminate the atmosphere and water sources. Manure or treated and diluted sewage sludge can be added to soil as a fertiliser.

- **Nitrates** are also produced by traditional and artificial fertilisers, and may be washed into water sources. Nitrates are probably responsible for diseases such as methaemoglobinaemia and stomach cancer. They are also responsible for the **eutrophication** of lakes and algal blooms, which cause the depletion of free oxygen in standing bodies of water, such as lakes, and the consequent suffocation of aquatic animals.

- Nitrates may be chemically removed from water by cation exchange, reverse osmosis, electrodialysis, distillation and biological processes, but these are expensive to implement.

- **Dangerous organic chemicals** (chlorinated hydrocarbons, notably PCBs, DDT and TBT) are manufactured for paints, plastics, adhesives, hydraulic fluids, electrical components, defoliants, pesticides and anti-fouling paints. The waste from such industrial processes is commonly dumped into watercourses. Such chemicals take a long time to biodegrade or break down, and therefore they can concentrate in food chains to poison animals and plants.

- **Heavy metals** (principally mercury, lead, arsenic, selenium, cobalt, copper and aluminium) form by weathering, but industrial processes discharge large quantities into watercourses that may be at toxic levels. The metals become concentrated at high trophic levels within food webs, and may have serious effects on life, e.g. causing brain damage and even death in animals. Lead piping for water resources, and the use of aluminium in cooking utensils, may further concentrate these metals in the human body. Aluminium may be a contributory factor in Alzheimer's and Guam disease. The sources of heavy metal pollutants can be traced using isotopic methods, and this may provide an effective means of tracing metal pollution to specific polluters.

- **Radioactive waste** poses a very serious threat to health but is dealt with more fully in Chapter 6.

- **Oil pollution** is a major environmental problem. It results from shipping accidents, offshore oil exploration, oil droplets from unburned fuel, combustion engines on land, deliberate discharges from ships (particularly during tank cleaning operations), natural seepage from the sea bed, and ecological terrorism. Oil pollution affects many ecosystems. Oil in water commonly coats fish gills, leading to suffocation and immobilisation of fish sperm. It also coats birds' feathers to impede flight and reduce insulation, leading to the death of birds, and it reduces the permeability of birds' eggs, reducing the numbers of offspring. It can also cause lipid pneumonia in mammals, leading to death. All this seriously disrupts food webs and devastates local ecosystems.

- **Clean-up technologies** for oil pollution include containing the spread of oil by using booms, dispersal and break-up of oil slicks using detergents, combustion, collecting the oil, and bacterial degradation.

- **Ground water** provides the largest available supply of fresh water and is important in sustaining the flow of rivers and is a vital component of the hydrological cycle. The dynamics of ground-water flow are important for resource management, and the mitigation of ground-water pollution and its remediation.

- The importance of water as a valuable resource causes international and regional conflicts, and governments are becoming increasingly aware of the need for long-term agreements between nations that must share common water resources.

Main learning hurdles

Hydrological cycle

The hydrological cycle was explained in Chapter 1, but it should be revised here so that the students can fully appreciate movement of polluted water within different environments.

Food chains and food webs

The dynamics of food chains should be described so that students with a poor biological background can appreciate how pollutants may become concentrated in higher trophic levels.

Eutrophication

The instructor should explain how nutrients can enter water bodies and cause increased productivity, which results in oxygen depletion. Many students may not fully appreciate the importance of oxygen in water, so the instructor should explain its essential role.

Ground water

Students without a geological background can be confused by the term 'ground water' because they do not understand what is meant by porosity, permeability, Darcy's Law, etc. The instructor should outline these basic concepts and if possible use rock samples to help the student to appreciate the characteristics of rocks, such as their porosity and permeability.

Key terms

Algal blooms; poly-chlorinated biphenyls (PCBs); Darcy's Law; desalination plant; desulfomaculum; desulfovibrio; dichloro-diphenyl-trichloro-ethane (DDT); eutrophication; food chain; food web; ground water; heavy metal; hydrological cycle; lipid pneumonia; nitrate; permeability; porosity; salinisation; self-purification.

Issues for group discussion

Discuss the different ways that ground water may be polluted and the means to reduce the effects.

The instructor should encourage the students to consider the different forms of pollution, ranging from radionuclides, sewage, metals, dangerous organic compounds, nitrates and oil to salt. The discussion should then consider the main ways that each of these groups of pollutants is dumped or spilled and is able to enter the ground. The students should consider the flow of pollutants in the ground and they should discuss the various methods of remediation. With respect to each type of pollutant, the discussion should continue by examining the costs and effectiveness of each of these methods.

Discuss how improved water quality can lead to increased development in poor countries.
The discussion might focus on the problems of water pollution in developing countries and the effects on health and consequently lower productivity. The discussion should include the methods of improving water quality and means to implement these methods.

Discuss the strategic importance of water resources.
Students should read Gleick (1994) to discuss the problems of conflicts that can occur in regions of sparse water, using the Middle East as an example. In addition, the students should also discuss the contents of Pearce (1994), which illustrate the types of destruction of water resources that have occurred during military conflicts.

Discuss the available options for dumping versus recycling redundant oil rigs/platforms.
This discussion could be initiated by outlining the background to the *Brent Spar* platform owned by the oil company Shell and at present languishing in Erfjord, near Stavanger, Norway. Students should be encouraged to explore the multiplicity of issues raised by the exploration and production of oil in marine environments from coastal to deeper, more offshore, waters, and the ultimate fate of the oil platforms. Currently (October 1996), one of the most favoured options for the disposal of the *Brent Spar* is to sell it to the giant construction company Amec for conversion into a marine leisure/viewing platform to be sited in Morecambe Bay, England.

Students should consider whether this possible option for the *Brent Spar* is merely a solution for one of the most controversial platforms, but that alternative solutions need to be found for the large number of other platforms that will require future disposal.

Also, students should discuss whether or not in the light of the environmental pressure group, Greenpeace, having dramatically overestimated the actual levels of toxic chemicals left in the *Brent Spar* (publically admitted by Greenpeace), this materially alters the underlying premise that the oceans should be a repository for dumping industrial waste. Finally, students should explore the issues associated with the considerable costs incurred in any clean-up and recycling of oil platforms.

Selected readings

Anderson, D.M. 1994. Red Tides. *Scientific American*, August, 52–8.
This is a good account of the biology of algal blooms and their toxic and environmental effects. It is useful for discussing the problems of marine pollution and changes in the marine ecosystem.

Burt, T. and Haycock, N. 1991. Farming and nitrate pollution. *Geography*, 76, 60–3.
This is a concise summary of the problems and effects of nitrate pollution resulting from agricultural activities.

Gleick, P.H. 1994. Water, war and peace in the Middle East. *Environment*, 36 (3), 6–15 and 35–42.
This is a useful article because it discusses the hydropolitical issues in the Middle East and provides recommendations for effective management to aid political stability in the region.

Hoffman, A.J. 1995. An easy rebirth at Love Canal. *Environment*, 37 (2), 4–9 and 25–31.
This is an interesting discussion of the effectiveness of environmental protection of water resources, illustrated by the Love Canal disaster.

Mitchell, R.B. 1995. Lessons from international oil pollution. *Environment*, 37 (4), 10–15 and 36–41.
This is a discussion of the international politics surrounding oil pollution. The article recommends methods to aid in effective legislation.

Nraigu, J.O. 1990. Global metal pollution: poisoning the biosphere? *Environment*, 32 (7), 6–12 and 28–33.
This is an interesting review of global metal pollution, with useful references and tables of data.

Nriagu, J.O. 1993. Legacy of mercury pollution. *Nature*, 363, 589–90.
This paper discusses the problems of mercury poisoning associated with past silver mining in South and Central America.

Pearce, F. 1995. Dead in the water. *New Scientist*, 4 February, 26–31.
This article discusses the problems of arsenic poisoning in ground-water resources.

Price, M. 1991. Water from the ground. *New Scientist*, 16 February, Inside Science, 42.
This is a discussion of the dynamics of ground water and the problems of pollution.

Pearce, F. 1994. Water in the war zone. *New Scientist*, 17 December, 13–14.
An interesting discussion of the problems associated with the destruction of water resources during wars.

Simonich, S.L. and Hites, R.A. 1995. Global distribution of persistent organochlorine compounds. *Science*, 269, 1851–4.
An assessment of the global distribution of organochlorine compounds. It uses the concentrations of pollutants in trees throughout the world. The paper also discusses the mechanisms involved in distributing these pollutants on a global scale.

Vivian, B. and Spiers, R. 1991. Waste disposal and the management of landfill sites. *Geography*, 76, 63–7.
An interesting article discussing the problems of waste management and landfill sites.

Textbooks

Domenico, P.A. and Schwartz, F.W. 1990. *Physical and Chemical Hydrogeology*. Chichester: John Wiley & Sons.
This is a comprehensive text on the physical and chemical aspects of ground water. The text is rather mathematical and is of interest to the specialist, but there are good chapters on contaminant hydrology and remediation.

Hinrichsen, D. 1990. *Our Common Seas: Coasts in Crisis*. London: Earthscan.
This book is based on UNEP data and describes the growing pressures on coastal ecosystems throughout the world. Case studies illustrate the local successes in protecting marine and coastal environments.

Kliot, N. 1993. *Water Resources and Conflict in the Middle East*. London: Routledge.
This book examines the hydrological, social, economic, political and legal issues in the Middle East. It shows how water shortages threaten the renewal of conflict and disruption in the Euphrates, Tigris, Nile and Jordan basins.

Mason, C.F. 1996. *Biology of Freshwater Pollution*, third edition. Harlow: Longman.
This comprehensive text provides an overview of aspects of fresh-water pollution. It is a useful reference source for all interested in environmental pollution.

Open University 1991. *Case Studies in Oceanography and Marine Affairs*. Oxford: Pergamon Press.
A well-illustrated textbook produced to support an Open University course on oceanography. It examines marine resources and activities, and the development and nature of international conventions of the sea, and it provides case studies on the Arctic Ocean and the Galapagos Islands. This book emphasises the complex interactions between the political, economic and environmental aspects of development within the marine realm.

Price, M. 1985. *Introducing Groundwater*. London: Chapman & Hall.
This is a useful introductory text to hydrogeology/hydrological aspects of ground water. The book is for non-specialists and uses minimal technical and mathematical formulae. There are thirteen chapters: water underground; water circulation; caverns and capillaries; soil water; ground water in motion; more about aquifers; springs and rivers; deserts and droughts; water wells; measurements and models; water quality; ground water, friend or foe?; some current problems.

Essay questions

1 Describe how the use of fertilisers may lead to the pollution of valuable water resources.

2 Evaluate the problems associated with the exploitation of water resources in EITHER small oceanic islands OR desert margins.

3 Describe the causes and effects of nitrate pollution in rivers and seas.

4 Discuss, with reference to specific examples, how international regional conflicts have developed because of the need to share water resources.

5 Describe the effects of oil pollution in coastal waters, and the available clean-up technologies.

6 Describe the ways in which toxic metals may become concentrated in water resources, and their possible effects on humans.

7 Discuss the pros and cons of using water meters to control the consumption of water in developed countries.

8 'The study of hydrogeology is essential for the management of water resources.' Discuss.

9 Evaluate the various legal methods used to control water pollution.

10 Describe the environmental and health problems associated with algal blooms.

11 With reference to the various classes of marine pollutants, explain why dispersal, dilution and degradation are effective in reducing environmental damage only for certain pollutants.

12 Assess the possible political problems associated with poor water management on an international scale.

Multiple-choice questions

Choose the best answer for each of the following questions.

1 The seas and oceans cover approximately:
   (a) 50 per cent of the Earth's surface
   (b) 66 per cent of the Earth's surface
   (c) 70 per cent of the Earth's surface
   (d) 75 per cent of the Earth's surface

2 The hydrological cycle describes the storage and movement of water through the:
   (a) hydrosphere
   (b) biosphere
   (c) ecosphere
   (d) lithosphere

3 Waste in water is degraded by microbes in water through a natural process called:
   (a) microbe cleansing
   (b) self-purification
   (c) eutrophication
   (d) biosynthesis

4 The oceans and seas comprise approximately:
   (a) 50 per cent of the total global water
   (b) 60 per cent of the total global water
   (c) 70 per cent of the total global water
   (d) 97 per cent of the total global water

5 Porosity describes the:
   (a) ability of rock to allow water to flow through it
   (b) connective nature of voids within a rock
   (c) ratio of voids to the volume of constituent rock particles
   (d) ratio of voids to total rock volume

6 Red tides are the result of:
   (a) algal blooms
   (b) iron contamination
   (c) oil pollution
   (d) nitrate pollution

7 Methaemoglobinaemia is a disease that is believed to be caused by water that has been contaminated by:
   (a) aluminium
   (b) lead
   (c) nitrates
   (d) DDT

8 Aluminium poisoning is believed to cause:
   (a) methaemoglobinaemia
   (b) cryptosporidia
   (c) Alzheimer's disease
   (d) leptospirosis

9 The transfer of food from one type of organism to another within a complex community of organisms is known as a:
   (a) food chain
   (b) food web
   (c) food mesh
   (d) food grid

10 Rocks that prevent water storage and allow water to be extracted for human use are referred to as:
    (a) aquifers
    (b) aquiclude
    (c) saturated
    (d) permeable

11 The level at which water stands within wells defines an imaginary surface, which is known as:
    (a) the water table
    (b) the potentiometric surface
    (c) the spring line
    (d) the hydrological surface

12 The law that describes the rate of flow of water through rock is called:
    (a) Darcy's Law
    (b) Boyle's Law
    (c) Dorris' Law
    (d) Charles' Law

13 The major source of lead pollution in water sources in South America is:
    (a) deforestation
    (b) mining activities
    (c) motor exhaust
    (d) industrial activities

14 Widespread arsenic poisoning in Bangladesh is the result of:
    (a) contamination of ground water due to industrial activities
    (b) the oxidation of iron sulphides in rocks related to depressed water tables
    (c) contamination of ground water and surface waters from landfill sites
    (d) industrial pollution of rivers and streams

15 Which of the following pollutants is most persistent in the oceans?
    (a) domestic sewage
    (b) oil
    (c) pesticides
    (d) toxic metals

16 The first person to recognise the connection between bad water quality and cholera epidemics was:
    (a) King Richard I of England
    (b) Joseph Chamberlain
    (c) Dr John Snow
    (d) Rachel Carson

17 DDT is the abbreviation for:
    (a) dichloro-diphenyl-trichloro-ethane
    (b) dichloro-diphenyl-trichloro
    (c) don't dump toxic waste
    (d) dibromochloromethane

18 Tributyltin (TBT) is used as a:
    (a) anti-fouling paint
    (b) pesticide
    (c) fertiliser
    (d) dispersant

19 Lipid pneumonia in mammals is a condition caused by inhaling:
    (a) lead
    (b) oil
    (c) DDT
    (d) bacteria

20 The major cause of nitrate contamination in water sources is the use of:
    (a) fertilisers
    (b) pesticides
    (c) liming agents
    (d) irrigation

21 The most common pollutant that contaminates domestic water sources is:
    (a) sewage and sludge
    (b) oil
    (c) toxic metals
    (d) pesticides

22 The drawing of salts and pollutants towards the surface of the ground due to evaporation is known as:
    (a) desalinisation
    (b) salinisation
    (c) saline evaporation
    (d) elevation

23 The phenomenon that involves the spreading out of a pollutant in ground water to form a plume along and perpendicular to the flow direction is known as:
    (a) plumation
    (b) circulation
    (c) dissipation
    (d) dispersion

24 One of the most sensitive inorganic tracers for increasing amounts of sewage in the oceans is:
    (a) silver
    (b) lead
    (c) gold
    (d) strontium

25 The potential environmental impacts of wastes are commonly expressed in terms of their:
- (a) biochemical oxygen demand
- (b) redox potential
- (c) pH
- (d) cation exchange capacity

Figure questions

1 Figure 5.6 illustrates the distribution of major oil slicks throughout the oceans. Answer the following questions.
- (a) Suggest reasons for this pattern.
- (b) Summarise in note form some basic, common statistics behind the five biggest oil slicks.
- (c) Explain why certain areas appear to be at greatest risk from oil spills.

2 Figure 5.7 shows the behaviour of oil released into the sea. Answer the following questions.
- (a) Suggest ways to accelerate the degradation of oil on the sea surface.
- (b) Describe the various ways in which oil may contaminate birds.
- (c) Under a variety of conditions, how long is oil likely to be persistent in the sea?

Short questions

1 What is activated sludge?

2 Describe the effects of mercury poisoning.

3 Outline the main causes of oil pollution.

4 Why do algal blooms occur?

5 List the main types of disease that are associated with water sources that have been contaminated with human excrement.

6 Describe the effects of nitrate pollution.

7 Define Darcy's Law.

8 Why is the phenomenon of dispersion important in ground-water studies?

9 What was Rachel Carson's major contribution to pollution awareness?

10 Describe how persistent the main types of pollutant are in water bodies.

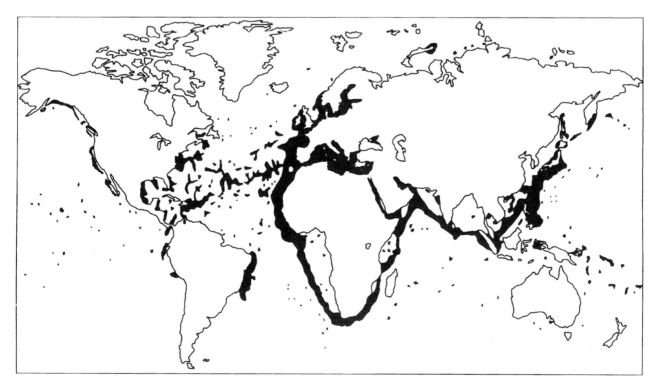

Figure 5.6 *Distribution of major oil slicks throughout the oceans (shown in black). Redrawn after Mysak and Lin (1990).*

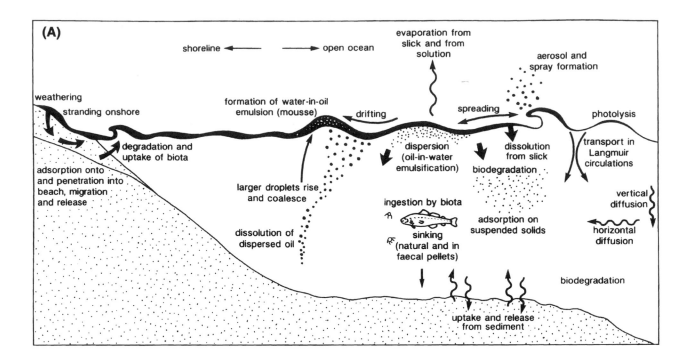

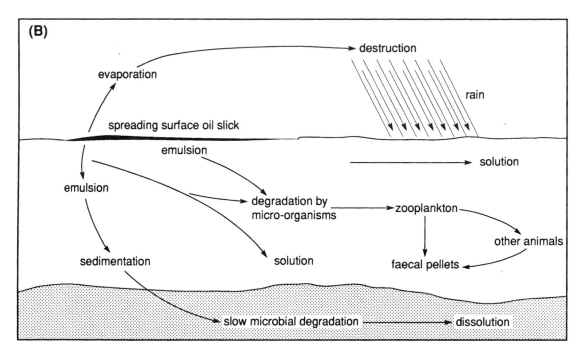

Figure 5.7 *(A) The behaviour of oil released into the sea. After Open University (1991). (B) The persistence of pollution in the oceans, its dispersal and degradation. Redrawn after Smith and Warr (1991).*

Answers to multiple-choice questions

1 c; 2 c; 3 b; 4 d; 5 d; 6 a; 7 c; 8 c; 9 b; 10 b; 11 b; 12 a; 13 b; 14 b; 15 d; 16 c; 17 a; 18 a; 19 b; 20 a; 21 a; 22 b; 23 d; 24 a; 25 a.

Answers to figure questions

1 (a) The pattern of oil pollution mimics the main shipping routes, particularly the routes taken by oil tankers. (b) See Table 1.1. (c) These are areas of major oil production and at crossroads of shipping routes.

2 (a) Methods of accelerating the degradation of oil on the sea surface include the use of emulsifiers and dispersants to break the oil into droplets to increase its surface area and aid bacterial degradation of the oil; the use of biological agents such as desulfovibrio and desulfomaculum, which use the oil as a food source; and fertiliser applications, which stimulate the growth of hydrocarbon-degrading micro-organisms within the intertidal zone. (b) Oil affects birds by coating their feathers, thereby reducing buoyancy and insulation. Lipid pneumonia and intestinal irritation can result. Oil may also affect birds' eggs, reducing the permeability of the egg, and oil pollution may cut down the food supply to birds as the fish life dies. (c) Oil may persist in the ocean for weeks to years, dependent upon how quickly it is dispersed and biologically degraded.

Answers to short questions

1 Sewage treatment uses activated sludge. It is a medium that contains living organisms that feed on solid sewage, encouraging its breakdown. Secondary treatment of sewage recycles the sludge.

2 If mercury accumulates in the body it can cause numbness of the limbs, speech impairment, loss of co-ordination and eventually death.

3 The main causes of oil pollution include shipping accidents; offshore oil exploration; droplets from unburned fuel; from combustion engines on land; deliberate discharges from ships (particularly during tank-cleaning operations); natural seepages from the sea bed; and ecological terrorism.

4 Algal blooms are the result of a proliferation of algae in water bodies as a result of changes in water chemistry and temperature. These changes may occur naturally due to processes such as upwelling or due to human activity such as pollution, particularly from the concentration of nitrates that wash from fertilised farmland.

5 The main types of disease associated with water sources that have been contaminated with human excrement include cholera, typhoid, cryptosporidia, leptospirosis and giardia.

6 Nitrates may be responsible for diseases such as methaemoglobinaemia (blue baby births) and stomach cancer. Nitrates may also be responsible for eutrophication of lakes and algal blooms, which cause the depletion of free oxygen in standing bodies of water, such as lakes, and the consequent suffocation of aquatic animals.

7 The rate of flow of water through a cylinder of rock is describe by Darcy's Law, such that:

$$Q = KA(h_L/l)$$

where Q is the rate of flow of water
K is a constant of proportionality, known as the hydraulic conductivity
A is the cross-sectional area of the cylinder
h_L/l is the hydraulic gradient

8 Dispersion is an important phenomenon in ground-water studies because it describes how a pollutant spreads out over time to form a plume along and perpendicular to the flow direction from the point of release. This phenomenon means that the flow will become diluted, and the period of time the pollutant will travel past a point will increase considerably. In this respect, poisons that have percolated into ground waters will remain there for long periods and will have widespread effects.

9 Rachel Carson's major contribution to pollution awareness was the publication of her book, *Silent Spring*, in 1962. The book described the effects of DDT on the environment and highlighted to the general public and policy-makers the threat of pollutants because they concentrate in higher trophic levels in food chains. Initially, many respected 'establishment' scientists appeared in the media to discredit Carson's predictions, for example claiming instead that rather than chemical pollution the bigger risk to humankind would be from insects. However, after a series of well-publicised major environmental disasters – e.g. the Minamata mercury discharge into a Japanese bay in 1959 (prior to Carson's book); the oil spills of the *Torrey Canyon*, *Amoco Cadiz* and *Exxon Valdez*, contaminated land at Love Canal in the USA; the Bhopal chemical cloud in India; Chernobyl and Three Mile Island – society has come to respect Carson's prescience. Also, Carson's book encouraged more concerted and focused scientific research, which eventually resulted in DDT being banned in many developed countries.

10 The most persistent types of pollutant in the hydrosphere are radioactive isotopes and toxic metals. These pollutants may remain for hundreds to thousands of years in water sources. Detergents, pesticides, PCBs and DDT may remain for years to many decades. Oil, however, usually degrades within a few years of the pollution and domestic sewage usually degrades within a few weeks to a few months.

Additional reference

Pearce, F. 1996. Dirty groundwater runs deep. *New Scientist*, 21 September, 16–17.

CHAPTER 6
Nuclear issues

Aims

- To examine the main types of nuclear issues, which include natural radiation; energy production and waste disposal; and nuclear weapons.

- To assess the pros and cons of nuclear energy and nuclear weapons.

- To examine the environmental effects of nuclear disasters.

- To evaluate the means of safely controlling the use of nuclear energy and weapons.

Key point summary

- The **Nuclear Age** began at Cambridge University, UK, in 1919, with Rutherford's experiments on the structure of the atom.

- **Nuclear fission** was first achieved by Hahn and Strassman in 1934.

- The development of the atomic bomb was sanctioned by Roosevelt in 1941, and the first two bombs were used against Japan in August 1945.

- **Treaties** signed to limit arms proliferation and nuclear testing include: the 1963 Limited Test Ban Treaty (LTBT); the 1968 Non-Proliferation Treaty (NPT); the 1972 Strategic Arms Limitation Treaty (SALT I); the 1974 Threshold Test Ban Treaty (TTBT); the 1976 Peaceful Nuclear Explosions Treaty (PNET); the 1979 Strategic Arms Limitation Treaty (SALT II); the 1987 Intermediate Nuclear Forces (INF) Treaty; the 1991 Strategic Arms Reduction Treaty (START); and the 1992 START 2.

- Policing nuclear missile deployment is difficult, and effective and adequate **verification mechanisms** are necessary in order to instil confidence in a treaty, and in order to check that no nation is cheating.

- Mechanisms for verification include seismic verification to monitor underground nuclear testing, remote sensing, scientific exchange programmes, and on-site inspections of nuclear installations.

- The collapse of the Soviet Union and the birth of the Commonwealth of Independent States and the Russian Federation may result in greater opportunities for significant arms reductions, although real concerns remain over the proliferation of nuclear weapons, and their pre-emptive use by politically unstable regimes in various war zones.

- **Radon** is a major component of the natural background radiation dose received by many people, and results from the decay of radioactive minerals in rocks.

- Tentative links have been drawn between cancer in humans and areas of high radon concentrations.

- Factors that concentrate radon in the environment include the geological setting and the bedrock type; building design and the materials used; water sources; and atmospheric conditions.

- Governments have commissioned surveys to identify areas of high risk and have provided recommendations to help to reduce the effects within the home.

- Nuclear waste results from the production of nuclear weapons and energy, medical products, and scientific research.

- Waste management in the nuclear power industry includes the **nuclear fuel cycle**.

- Prior to 1970 for the USA and 1982 for Europe, waste could be dumped at sea, but since 1975 all dumping has had to comply with the **London Dumping Convention**.

- The half-life and toxicity of radioactive chemicals vary greatly, a factor that determines the risk and magnitude of any potential contamination.

- The transport of nuclear waste is hazardous, and accidents may affect very large areas for extremely long periods.

- Acceptable underground nuclear waste disposal relies on many factors, including the geological setting and bedrock geology, ground-water movement, the nature of the containers in which the waste is sealed, and its monitorability and retrievability.

- Nuclear accidents at power stations have occurred, for a variety of reasons, but all involving some degree of human error, and include Three Mile Island in 1979, Chernobyl in 1986 and Tomsk-7 in 1993. These accidents serve to highlight the potential dangers, clean-up problems and costs, and the environmental damage.

Main learning hurdles

Radioactive decay

The instructor should explain the basic chemistry and physics involved in the understanding of radioactive decay. This involves the structure of the atom, isotopes, and the nature of alpha, beta and gamma decays. A half-life should also be explained using the appropriate decay curves. This will enable students to understand the notations, the nature of radioactive radiation and how persistent radioactive material is in the environment.

Underground disposal of nuclear waste

It is commonly difficult for students with a weak geological background to appreciate fully the problems of disposing of nuclear waste underground. The instructor should spend a little time explaining the nature of bedrock, for example permeability and porosity, and the nature of earth movements, for example a revision of plate tectonic theory.

Key terms

Acute radiation syndrome (ARS); alpha particle; background radiation; beta activity; half-life; meltdown; mutually assured destruction (MAD); nuclear fission; nuclear winter; phosphate nodule; plutonium; radioactive decay; radon gas; röentgen.

Issues for group discussion

Discuss the assertion that nuclear weapons are no longer a threat because there has been no nuclear war since the invention of the atom bomb, particularly in the post-Cold War period.

The discussion might focus on the various arms treaties and the threat of the newly emerging countries in the former Soviet Union and other countries that have the potential to develop weapons in the future. Students should discuss the likely future conflicts, particularly as a result of environmental degradation.

Discuss the various pros and cons of disposing of nuclear waste in the ground.

The students should discuss the various types of waste and their longevity. They should also discuss the uncertainty associated with the underground disposal of nuclear waste and emphasise the instabilities within the geological environment. In addition, they should consider the alternatives to underground waste disposal.

Selected readings

Furth, H.P. 1995. Fusion. *Scientific American*, September, 174–7.
This is a good account of the theory of nuclear fusion.

Hassard, J. 1992. Arms and the ban. *New Scientist*, 28 November, 38–41.
This is a useful, easy to read article on arms reduction.

Peto, J. and Darby, S. 1994. Radon risk reassessed. *Nature*, 368, 97–8.
This is an authoritative assessment of radon hazard.

Williams, N. 1995. Chernobyl: life abounds without people. *Science*, 269, 304.
This paper discusses some of the ecological effects of the nuclear disaster at Chernobyl.

Textbooks

Berkhout, F. 1991. *Radioactive Waste: Politics and Technology*. London: Routledge.
This is a companion book for students of environmental studies, geography and public administration. The book focuses on radioactive waste management and disposal policies in three European countries

– the UK, Germany and Sweden. It presents a detailed historical account of the policy processes in these three countries, and evaluates the theoretical and public policy implications. The book's particular strength is in its comparative approach, and the way in which it sets out the issue of radioactive waste management at the centre of the current debate about nuclear power, the environment and society.

Medvedev, Z.A. 1990. *The Legacy of Chernobyl.* London: W.W. Norton.
This is an interesting account of the catastrophe at Chernobyl. It provides a post mortem of the events and describes the environmental effects, including the impact on agriculture, the health impact in the former USSR and the global impacts. Furthermore, it discusses the Soviet nuclear energy programme and analyses the history of nuclear accidents in the former USSR. This is followed by a consideration of the future of nuclear power in Russia.

Newhouse, J. 1989. *The Nuclear Age: From Hiroshima to Star Wars.* London: Michael Joseph.

This book is a history of the post-Second World War events and personalities behind the single most important issue of the past 50 years – nuclear weapons and the nuclear arms race. This book is well researched and highly readable. It tells a dramatic story of confrontation and rapprochement, of scientific and technological advance, of diplomatic wrangling and blatant shows of military strength, through the Cold War and the Cuban Missile Crisis, to the Star Wars programme, and the more recent US–former Soviet Union arms agreements. It is an invaluable account of the Nuclear Age.

Sheehan, M.J. 1988. *Arms Control: Theory and Practice.* Oxford: Blackwell.
This book provides a useful analysis of the origins and development of arms control, and the issues that underpin arms control. The problems of verifying treaties, and the political context in which arms control negotiations, both domestic and international, are considered. This is a useful supplementary book for any student interested in understanding something of the complexity of arms control issues.

Essay questions

1 'Nuclear energy provides a clean alternative to fossil fuels such as coal, oil and gas.' Discuss.

2 'Nuclear weapons and the threat they pose have kept humankind from a global war.' Discuss.

3 Describe the environmental effects produced by natural and artificial radioactivity.

4 Outline the treaties and conventions to control nuclear tests and the proliferation of nuclear arms.

5 What considerations should be taken into account for the safe disposal of low- and intermediate-level radioactive waste?

6 Describe the various processes to aid in the verification of nuclear treaties.

7 Discuss the arguments for and against underground nuclear testing on atolls.

8 'Now that the Cold War is over, there is no need for nuclear weapons.' Discuss.

9 Describe the immediate after-effects and the long-term consequences of the Chernobyl accident.

10 Describe the effects and various ways of reducing the radon hazard in buildings.

11 Discuss the environmental and political issues surrounding the 1995–6 French nuclear tests in the South Pacific.

12 Describe the possible effects of a 'nuclear winter'.

Multiple-choice questions

Choose the best answer for each of the following questions.

1 An alpha particle is emitted from the:
   (a) nucleus of an atom during radioactive decay
   (b) shell of an atom during radioactive decay
   (c) total destruction of the nucleus of an atom during radioactive decay
   (d) total destruction of an atom during radioactive decay

2 Symptoms resulting from intensive irradiation of the body, including nausea, vomiting, abdominal pain, fever, dehydration, loss of hair, infection, haemorrhage, damage to bone marrow and cancers, is known as:
   (a) radiation sickness
   (b) radiation cancer
   (c) acute radiation syndrome
   (d) nuclear syndrome

3 Which of the following units is used to measure radioactivity?
   (a) radons
   (b) joules
   (c) oppenheimers
   (d) becquerels

4 The splitting of larger atoms into smaller atoms is known as:
   (a) decomposition
   (b) disintegration
   (c) fission
   (d) fusion

5 Isotopes are elements that have the same number of:
   (a) neutrons in the nucleus but have different numbers of electrons in the outer shell
   (b) protons in the nucleus but have different numbers of neutrons
   (c) protons in the nucleus but have different numbers of electrons in the outer shell
   (d) neutrons in the nucleus but have different numbers of protons

6 The average amount of radon in household air in Britain is approximately:
   (a) 20 Bq m^{-3}
   (b) 200 Bq m^{-3}
   (c) 2,000 Bq m^{-3}
   (d) 20,000 Bq m^{-3}

7 Nuclear fission was first achieved by:
   (a) Rutherford
   (b) Hahn and Strassman
   (c) Oppenheimer
   (d) Currie

8 The first treaty to limit arms proliferation and nuclear testing was called:
   (a) Limited Test Ban Treaty
   (b) Non-Proliferation Treaty
   (c) Strategic Arms Limitation Treaty
   (d) Threshold Test Ban Treaty

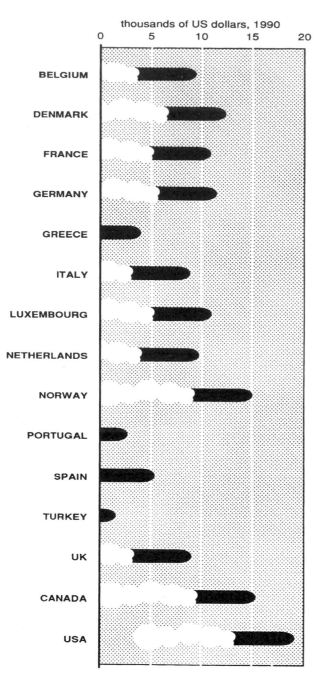

thousands of US dollars, 1990

BELGIUM
DENMARK
FRANCE
GERMANY
GREECE
ITALY
LUXEMBOURG
NETHERLANDS
NORWAY
PORTUGAL
SPAIN
TURKEY
UK
CANADA
USA

Figure 6.1 *Comparison between the military expenditure per capita of the top 15 developed countries. From* The Environmental Guardian *(11 February 1992).*

9 Which of the following radioactive isotopes has the shortest half-life?
   (a) ^{238}U
   (b) ^{220}Rn
   (c) ^{137}Cs
   (d) ^{90}Sr

10 Which of the following US Presidents sanctioned the construction of the first atomic bomb?
   (a) Kennedy
   (b) Roosevelt
   (c) Truman
   (d) Reagan

Figure questions

1 Figure 6.1 shows the comparison between military expenditure per capita of the top 15 developed countries. Answer the following questions.
   (a) Select three countries with very different military spending patterns and discuss why they might have widely differing military expenditure.
   (b) Find out what the most up-to-date military spending patterns are for the USA, China and Japan.
   (c) Find out what the per capita military spending is for a developing nation not represented in Figure 6.1.

2 Figure 6.2 is an idealised diagram showing (a) alpha-decay and (b) beta-decay processes. Answer the following questions.
   (a) Describe alpha-decay and give an example.
   (b) Describe beta-decay and give an example.
   (c) What is the mathematical expression to describe the half-life of a radioactive element?

3 Figure 6.6 shows the ways in which radon gas enters the home. Answer the following questions.
   (a) Describe a radioactive decay path that leads to the creation of radon gas.
   (b) What are the potential health risks to humans caused by radon gas?
   (c) What are some of the most common rock types that are associated with the production of radon gas?

Short questions

1 What is ARS?

2 Describe the effects of a nuclear winter.

3 What is a radioactive isotope?

4 List the various ways that radon may enter homes.

5 What is a pozzolan material?

6 List the main treaties that have been signed to help to limit arms proliferation and nuclear testing.

7 Explain the meaning of NIMBY.

8 Describe the methods that can be used to verify nuclear treaties.

9 Define the term half-life.

10 List the major nuclear accidents.

Answers to multiple-choice questions

1 a; 2 c; 3 d; 4 c; 5 b; 6 a; 7 b; 8 a; 9 b; 10 b.

Answers to figure questions

1 Answers to this question are not provided as they involve students doing their own research.

2 (a) A radioisotope undergoes alpha-decay when an alpha particle forms. An alpha particle comprises two protons and two neutrons (essentially a helium nucleus), which results in a decrease of both the atomic number and atomic mass number of the radioisotope. (b) A radioisotope undergoes beta-decay when a beta particle forms. A beta particle forms when one of the protons or neutrons in the nucleus of a radioisotope spontaneously changes into a proton, which remains in the nucleus while an electron (beta particle) is emitted. A proton may change into a neutron, or a neutron may be transformed into a proton, and as a result of this process another particle, a neutrino, is also ejected. A neutrino is a particle with no rest mass. In this case the atomic number is reduced. (c) The mathematical expressions to describe the half-life of a radioactive element is given by:

$$N = N_0 e^{-\lambda t} \text{ and } T_{1/2} = \ln 2/\lambda$$

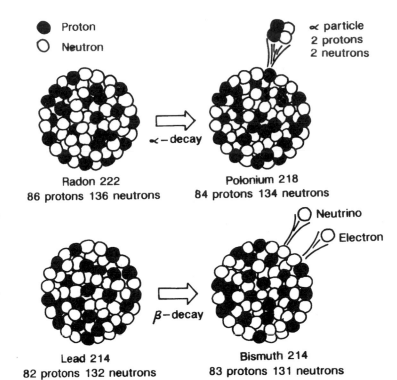

Figure 6.2 *Idealised diagrams showing (A) α-decay and (B) β-decay processes. Redrawn after Botkin and Keller (1995).*

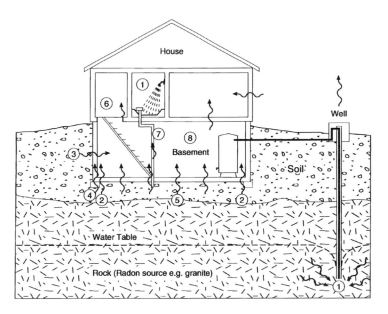

Figure 6.6 *Ways in which radon can enter homes. via 1 = ground water and domestic water for drinking, cooking and bathing; 2 = construction joints; 3 = cracks in walls; 4 = cavities in walls; 5 = cracks in solid floors; 6 = cracks in suspended floors; 7 = service pipes; 8 = radon concentrating in basements.*

where N = number of observed radioactive atoms; N_o = the initial number of radioactive atoms at time zero; λ = decay constant; t = time; $T_{1/2}$ = half life.

3 (a) Radon-222 is produced in the decay chain starting with uranium-238. Uranium-238 decays to thorium-234 (by α-decay and γ-decay) to protactinium-234 (by β-decay and γ-decay) to uranium-234 (by β-decay and γ-decay) to thorium-230 (by α-decay and γ-decay) to radium-226 (by α-decay and γ-decay) to Radon-222 (by α-decay and γ-decay), which in turn decays to polonium-218 (by α-decay). (b) Radon is tentatively linked to high incidence of cancers such as lung cancer and leukaemia. (c) Granite, shales, ironstones and phosphate nodules are among the common rocks that produce radon gas.

Answers to short questions

1 ARS is an abbreviation for acute radiation syndrome, which results from intense irradiation of the body. Symptoms include nausea, vomiting, abdominal pain, fever, dehydration, loss of hair, infection, haemorrhage, damage to bone marrow and cancers, notably leukaemia and breast cancer.

2 Multiple nuclear explosions, such as would be associated with a nuclear war, are likely to result in a nuclear winter. Severe deterioration of climate will result. Great fires and winds are formed by the explosions, and large quantities of smoke and dust are likely to be ejected into the upper atmosphere, where they remain for periods of months to years. This would cause prolonged darkness and reduced incoming solar radiation, resulting in extreme cooling of the Earth's surface, possibly by between 15 °C and 25 °C.

3 Isotopes are different forms of an element with similar chemical properties by virtue of atomic weight, that is, they have the same number of protons in the nucleus and the same number of electrons in their orbits, but isotopes have a different number of neutrons in their nucleus. A radioactive isotope is an isotope of a chemical element that is naturally unstable, and tends to become more stable by the emission of radioactive particles, for example alpha, beta and gamma radiation.

4 Radon may enter homes via ground water, in bathing and drinking water; through construction joints; through cracks in the walls; through cavities in the walls; through cracks in solid floors; through cracks in suspended floors; and through service pipes.

5 A pozzolan material is a substance that acts like a cement, and contains silicates and aluminosilicates that react with lime and water to form insoluble compounds. It is used in the disposal of toxic waste.

6 The main treaties that were signed to limit arms proliferation and nuclear testing include: the 1963 Limited Test Ban Treaty (LTBT); the 1968 Non-Proliferation Treaty (NPT); the 1972 Strategic Arms Limitation Treaty (SALT I); the 1974 Threshold Test Ban Treaty (TTBT); the 1976 Peaceful Nuclear Explosions Treaty (PNET); the 1979 Strategic Arms Limitation Treaty (SALT II); the 1987 Intermediate Nuclear Forces (INF) Treaty; the 1991 Strategic Arms Reduction Treaty (START); and the 1992 START 2.

7 NIMBY is an acronym for 'not in my back yard'. It epitomises the widely held public attitude that undesirable facilities (nuclear power plants, waste disposal utilities, chemical plants, etc.) should not be sited near their homes and/or workplaces but rather near somebody else.

8 Verification mechanisms include seismic verification to monitor underground nuclear explosions; remote sensing; scientific exchange programmes; and on-site inspections of nuclear installations.

9 A half-life is the time required for 50 per cent of the atoms of a radioactive isotope to decay to a different element/group of elements, with the associated emission of various subatomic particles and the release of energy.

10 Major nuclear accidents include: Three Mile Island in 1979; Chernobyl in 1986; and Tomsk-7 in 1993.

Additional references

Carrigan, C.R., Heinte, R.A., Hudson, G.B., Nitao, J.J. and Zucca, J.J. 1996. Trace gas emissions on geological faults as indicators of underground nuclear testing. *Nature*, 382, 528–31.

Williams, P. and Woessner, P.N. 1996. The real threat of nuclear smuggling. *Scientific American*, January issue, 26–30.

Aims

- To examine the needs and the main types of energy sources.

- To assess the pros and cons of different energy sources.

- To examine the environmental effects of different types of energy.

- To evaluate national and international energy strategies.

Key point summary

- Industrialisation, population growth, increased living standards, concerns about pollution and the serious depletion of many traditional energy resources such as fossil fuels have increased the need to develop renewable and clean energy.

- Most of the global energy supply comes from finite resources (77 per cent from essentially fossil fuels), from renewables (18 per cent from hydropower, wood, crop waste, dung and wind), and nuclear energy (*c.* 5 per cent).

- Conventional fossil fuels (coal, oil and gas) have low reserve lifetimes (e.g. world coal = *c.* 200–400 years, world oil = *c.* 56 years), but these figures have a large uncertainty attached to them because proven reserve estimates are modified by improved technologies of exploration and recovery, together with fluctuations in world commodity prices.

- Extraction and use of fuels causes many environmental problems, including open-cast mining and piling of waste, which scar the landscape. Leakages of oil and gas from pipelines and installations also cause environmental pollution problems.

- In the fossil fuel sector, environmental awareness has led to improved, cleaner technologies. Clean coal technologies, for example, are being developed, which include fluidised-bed combustion, flue-gas desulphurisation, gas-fired combined cycle gas turbines, gasification, and British Coal's Topping Cycle.

- To reduce atmospheric pollution, many governments in the developed countries have agreed to reduce their emissions by target dates, but such agreements have yet to make any headway in some of the major polluters, such as China and the former Soviet Union, two of the largest users of coal.

- Nuclear energy, produced by nuclear fission, has been developed since the Second World War. The industry has encountered many problems during this time, including several nuclear accidents and the problems associated with the safe disposal of nuclear waste. These problems have resulted in some countries deciding to reduce their dependence on nuclear power, and even mothball nuclear plants. Decommissioning nuclear power stations also has an environmental and cost problem.

- In the future, hydrogen energy, a form of chemical fuel, which could be easily transported and readily stored, may provide a major part of the world's energy requirements.

- Renewable energy resources include hydroelectricity, wind power, tidal energy, wave power, solar energy, geothermal energy and biomass energy. Technologies are being developed to increase the efficiencies of these fuels.

- Environmental problems, however, are associated with renewables, including the disruption of ecosystems, visual pollution and the current high production costs. Also, established energy businesses and cartels operate to minimise the potential competition for as long as possible.

- Cumulatively, private vehicles use large amounts of energy and contribute to poor air quality and other forms of environmental pollution.

- Improved public transportation systems and better urban planning could lead to a considerable reduction in pollution.

- Liquid bio-fuels, including bio-diesel and bio-ethanol, used as substitutes for conventional petrol and diesel, could provide environmentally more friendly alternatives. Although the technology to use these substitutes is sufficiently developed, the hydrocarbon and allied chemical industries have strong vested interests in slowing the introduction of bio-fuels, and without a distortion of the traditional fuel markets through favourable tax incentives and subsidies it seems unlikely that bio-fuels will become important until the middle of the twenty-first century.

- Attempts to reduce fuel consumption and improve air quality, including a reduction in the emissions of greenhouse gases caused by the combustion of fossil fuels, have involved proposals for the introduction of a carbon tax, increased use of energy-efficient technologies, energy conservation, and a greater reliance on renewable energy resources.

Main learning burdles

Units of measurement

The instructor should spend a little time explaining the differences between energy, power and work, emphasising the different types of units that are used for quantification. This will allow the students to compare the effectiveness and efficiency of the different types of energy discussed in this chapter.

Key terms

Anthracite; billion barrels of oil-equivalent (BBOE); bio-diesel; bio-fuel; biomass energy; bituminous coal; carbon/energy tax; carcinogenic; demand-side management (DSM); gasohol; geothermal energy; geothermal gradient; hydroelectric dam; hydrogen energy; lignite; low emission vehicle (LEV); natural gas; pneumoconiosis; rape methyl ester (RME); smart material; transitional low-emission vehicle (TLEV); ultra-low emission vehicle (ULEV); wind turbine; zero-emission vehicle (ZEV).

Issues for group discussion

Discuss the various ways in which city planners can help to increase energy efficiency.
The discussion should focus on how improved city planning can reduce the use of vehicles and improve life quality. The students should describe the method of building design that improves insulation, solar energy, etc., as well as the realities and problems of undertaking these improvements.

Discuss the problems associated with large-scale energy developments such as HEP, tidal barrages and nuclear power stations.
The discussion should consider the relative merits of each of these schemes as well as the associated environmental problems. The students must read Raphals (1992) and Pearce (1995) before the discussion.

Assess the recent technological advances that promise to provide cheap renewable energy sources.
The discussion should concentrate on developments in solar energy, hydrogen energy and liquid bio-fuel technology. The discussion should also consider if it is realistic to reduce the use of conventional fuel in favour of these new types of fuel.

Selected readings

Feldman, D.L. 1995. Revisiting the energy crisis: how far have we come? *Environment*, 37 (4), 16–20 and 42–4.
This is a very useful article that discusses energy needs.

Goldemberg, J. 1995. Energy needs in developing countries and sustainability. *Science*, 269, 1058–9.
This is a very authoritative paper on the energy needs of developing countries.

Hoagland, W. 1995. Solar energy. *Scientific American*, September, 170–3.
This is a useful review paper on solar energy technology.

Pearce, F. 1995. The biggest dam in the world. *New Scientist*, 28 January, 25–9.
This is an informative article discussing the environmental consequences of the construction of the Three Gorges Dam on the Yangtze River, China.

Raphals, P. 1992. The hidden cost of Canada's cheap power. *New Scientist*, 15 February, 50–4.

This article discusses the environmental problems associated with flooding large areas of northwest Canada for the construction of a reservoir and dam for hydroelectric power generation.

Textbooks

Blunden, J. and Reddish, A. (eds) 1991. *Energy, Resources and Environment*. London: Hodder & Stoughton.

This is a well-written and well-illustrated textbook, dealing with the nature of energy and mineral resources, their extraction, refining and disposal, and the environmental problems associated with obtaining these resources. The book explores the possible alternatives to conventional energy resources, including substitution and recycling, energy conservation, solar energy, and wind and water energy. It also considers the political implications of using such alternative energy resources. There is a whole chapter devoted to an examination of the politics associated with the disposal of radioactive waste.

Cassedy, E.S. and Grossman, P.Z. 1990. *Introduction to Energy Resources, Technology, and Society*. Cambridge: Cambridge University Press.

In this textbook, Cassedy and Grossman explore energy issues, and examine the benefits and problems associated with energy technology. The book is in three parts: Part I, dealing with energy resources and technology; Part II, examining power generation, the technology and its effects; and Part III, evaluating energy technology in the future. This book will prove useful as supplementary reading for students with a general interest in energy issues, particularly energy technology.

Johansson, T.D., Kelly, H., Reddy, A.K.N. and Williams, R.H. (eds) 1993. *Renewable Energy for Fuels and Electricity*. London: Earthscan in association with the United Nations.

This is a comprehensive and authoritative assessment of the technical and commercial potential for renewable forms of energy.

Essay questions

1 'Hydroelectric and tidal power stations provide cheap renewable energy, but the construction of such plants and their continued use results in serious environmental problems in estuaries and associated coastal waters'. Discuss.

2 Discuss the viewpoint that the production of electricity from nuclear power plants is more environmentally friendly than power produced by conventional fossil fuel plants.

3 The lifetime for global oil reserves is probably about 56 years. How reliable are such estimates, and what is the nature of the assumptions that go into such figures? If you were in charge of a nation's energy policy, what would you do about this prediction?

4 Explain the reasons why alternative renewable energy resources have not been utilised to a greater extent than at present.

5 What is a carbon and energy tax? Do you think it could have a positive impact on environmental pollution and help in energy conservation? What alternatives, if any, exist?

6 Describe the hazards associated with the construction of large reservoirs built for hydroelectric generation.

7 Suggest various means of reducing energy consumption in large cities.

8 Assess the relative merits of nuclear and coal-fired power stations.

9 Evaluate the status of liquid bio-fuels as acceptable and useful alternatives to conventional fuels.

10 'One of most important methods of helping to eradicate poverty in developing countries is the installation of electricity and the supply of cheap fuel.' Discuss.

11 Suggest effective ways that energy consumption may be reduced in urban areas.

12 Compare the characteristics of energy consumption and the associated environmental problems in developed and developing countries.

Multiple-choice questions

Choose the best answer for each of the following questions.

1 Which of the following formulae is benzene?
(a) C_6H_6
(b) $C_6H_6O_2$
(c) $C_6H_6CO_2$
(d) C_6H_6O

2 Which of the following is the highest-quality coal?
(a) bituminous coal
(b) lignite
(c) anthracite
(d) charcoal

3 A mixture of H_2 and CO used to synthesise liquid fuels such as pure hydrogen, methanol and gasoline is called:
(a) syngas
(b) gasohol
(c) bio-fuel
(d) hydromoncarbonate

4 Rape methyl ester is produced from:
(a) oilseed
(b) a mixture of alcohol and petroleum
(c) sugar
(d) fractional distillation of petroleum

5 A device capable of changing solar energy directly into electricity is called a:
(a) solar cell
(b) photocell
(c) solar-photocell
(d) photovoltaic cell

6 A blend of alcohol and conventional petroleum products is called:
(a) bio-fuel
(b) gasohol
(c) glycerine
(d) rape methyl ester

7 Which of the following hydrocarbon(s) are present in crude oil?
(a) kerogen
(b) methane
(c) benzene
(d) a, b and c

8 Which of the following methods of electricity generation produces the largest quantities of CO_2 emissions?
  (a) nuclear
  (b) hydroelectric
  (c) wind
  (d) tidal

9 Which of the following fossil fuel power plant systems is the most efficient?
  (a) pulverised fuel and flue-gas desulphurisation and low NO_x burners
  (b) circulating fluidised-bed combustion
  (c) pressurised fluidised-bed combustion
  (d) integrated gasification combined cycle

10 What is the current estimate for the world's coal reserve lifetimes?
  (a) 50 to 100 years
  (b) 100 to 200 years
  (c) 200 to 400 years
  (d) >400 years

Figure questions

1 Figure 7.9 is the Rotterdam product prices for premium gasoline, gas oil and heavy fuel oil for the years 1975 to 1992 inclusive (redrawn from British Petroleum's Statistical Review of World Energy, June 1993). Answer the following questions.
  (a) In the early 1980s, why did the Iran–Iraq war affect oil prices?
  (b) How meaningful do you think it is to try and predict possible future trends in the price of crude oil? Why do economists endeavour to make such predictions?
  (c) Why did the 1991 Gulf War affect oil prices?

2 Figure 7.10 compares two World Energy Council long-range energy scenarios. Answer the following questions.
  (a) Summarise the difference between the scenarios involving a continuation of present energy policy trends and those that are ecologically driven.

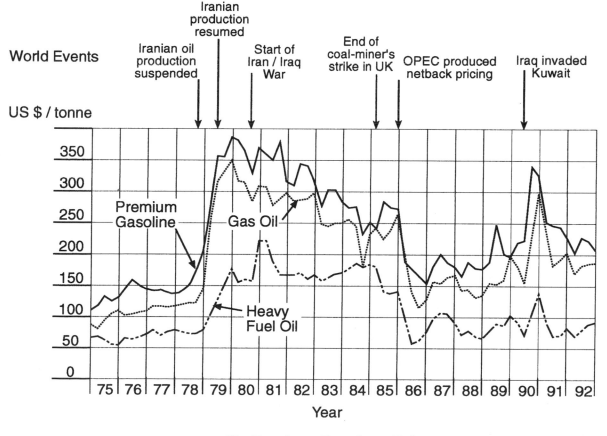

Rotterdam Product Prices

Figure 7.9 *Rotterdam product prices for premium gasoline, gas oil and heavy fuel oil between 1975 and 1992. Source: British Petroleum (1993).*

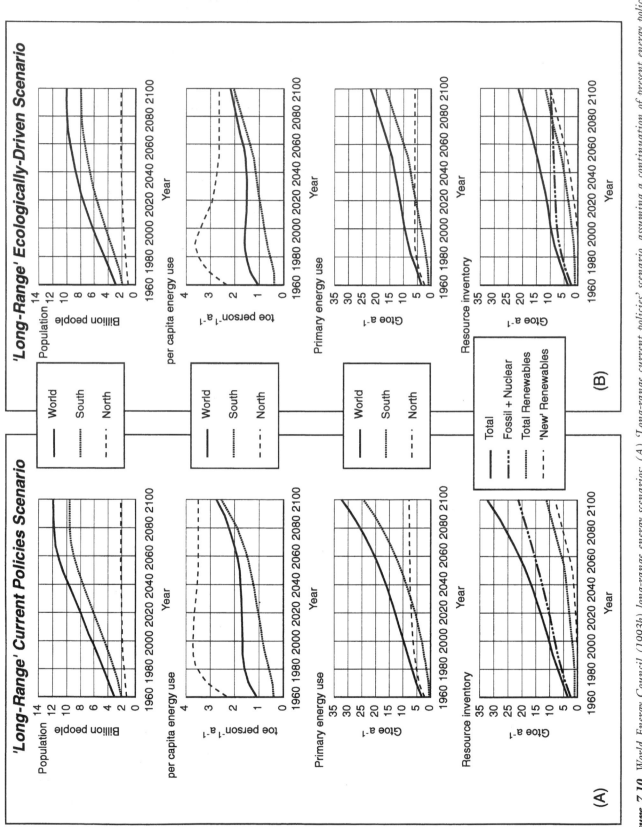

Figure 7.10 *World Energy Council (1993b) long-range energy scenarios: (A) 'Long-range current policies' scenario, assuming a continuation of present energy policy trends; (B) 'Long-range ecologically driven' scenario, assuming a higher priority is given to environmental considerations.*

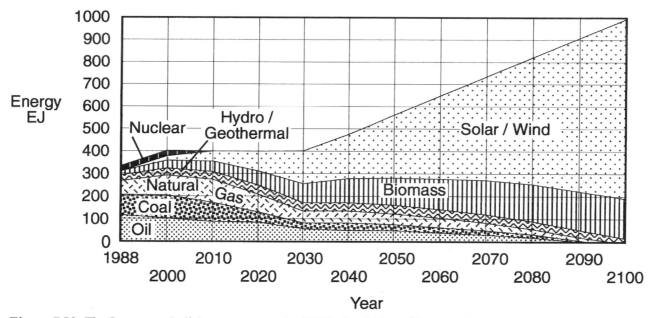

Figure 7.12 *The Greenpeace fossil-free energy scenario (FFES), in which world energy demand increases by a factor of approximately 2.5 by 2100. By this date, all fossil and nuclear fuels are envisaged as phased out and replaced by a mix of solar, hydro-, wind, biomass and geothermal energy sources. Redrawn after Lazarus et al. (1993).*

(b) What do you think are the uncertainties in the ecologically driven energy scenario?

3 Figure 7.12 is the Greenpeace Fossil-Free Energy Scenario (FFES). Answer the following questions.
  (a) Discuss the likelihood of the phasing-out of a nuclear component to the world energy production within the time-frame shown.
  (b) To what extent does Greenpeace rely on energy conservation measures versus using alternative energy sources to conventional fossil fuels and nuclear energy in its FFES?

Short questions

1   What is atmospheric fluidised-bed combustion?

2   List the main types of renewable energy resources.

3   What is a liquefied petroleum gas?

4   Describe the various attempts to reduce fuel consumption to improve air quality.

5   What is made from oilseed rape?

6   Estimate the reserve lifetimes for conventional fossil fuels.

7   List the main pollutants that result from motor vehicles.

8   What is the ALTENER plan?

9   Describe the environmental problems associated with the construction of large dams.

10 What is the difference between nuclear fission and nuclear fusion?

Answers to multiple-choice questions

1 a; 2 c; 3 a; 4 a; 5 d; 6 b; 7 d; 8 a; 9 a; 10 c.

Answers to figure questions

1 (a) Events in the Middle East, particularly the suspension of Iranian oil production and the Iran–Iraq war. The actual and perceived global shortage of crude oil supplies to the developed, industrialised nations caused oil prices to rise sharply. (b) In the absence of global crises that significantly affect the economies of oil-producing nations, there is a tendency towards greater accuracy in predicting the price of crude oil on the commodities markets. Regional conflicts (e.g. wars), however, can take the

world commodities markets by surprise and result in large discrepancies between economic predictions and prices. Despite the inherent problems in predicting oil prices, economists and traders on the commodities markets endeavour to make accurate predictions about crude oil prices as they underpin energy production and manufacturing in most countries. (c) The invasion of Kuwait by Iraq stopped the production of crude oil by one of the world's major suppliers. The disruption to oil supplies was effectively caused by the Gulf War itself, but after the war ended they were indirectly affected by the Iraqi action of setting fire to the oil wells.

2 (a) The principal difference between both long-range energy scenarios is that one is for current policies (i.e. no change or business-as-usual) while the other is for an ecologically driven set of energy policies. The WEC current policies scenario assumes that total world primary energy demand increases from 8.8 Gtoe in 1990 to approximately 32 Gtoe by the year 2100, and that in the year 2100 renewable energy resources, both 'traditional' and 'new', would provide about 33 per cent of the energy. In the WEC ecologically driven scenario, the assumption is that total primary energy demand will increase to approximately 22 Gtoe by the end of the twenty-first century, and that renewables will then provide about 12 Gtoe, accounting for just over 50 per cent. (b) The ecologically driven energy scenario includes uncertainties about the energy efficiency of existing, developmental and new technologies, unforeseen environmental damage from perceived 'clean' and 'safe' energy technologies that force a radical redefinition of energy policies, and the increased reliance on nuclear energy.

3 (a) After setting out selected world energy scenarios that both include and exclude a projected component of nuclear energy (e.g. WEC versus Greenpeace scenarios), the student should discuss the realistic likelihood of meeting future energy demand well into the next century without using nuclear energy with present and projected consumption rates. There may be a discussion of nuclear fusion versus nuclear fission for meeting energy demand. There should be a discussion of the perceived undesirability among many people of nuclear waste, e.g. immediate risks from nuclear accidents and bequeathing nuclear waste to the custody of future generations. Ideally, the essay should represent a balanced evaluation of the pros and cons of nuclear power, irrespective of the adopted attitude about nuclear energy. (b) The Greenpeace fossil-free (and nuclear-free) energy

scenario (FFES), developed for Greenpeace International by the Stockholm Environmental Institute's Boston centre (published as Lazarus *et al.* 1993), is one of the more optimistic models for the use of renewable energy. This model utilises conventional United Nations projections for world population to increase to more than 11 billion by the year 2100, with world GDP showing a fourteen-fold increase between 1988 and the year 2100. The FFES assumes that world primary energy demand will rise to approximately 1,000 EJ, equivalent to *c.* 24,000 Mtoe (by a factor of about 2.5), by the end of the twenty-first century. By this time, all fossil fuels and nuclear energy will have been phased out to be replaced by a mix of hydro-, solar, wind, biomass and geothermal energy (renewables). Improvements to energy efficiency (utilising market or near-market energy technologies) form a major part of the Greenpeace FFES, including the co-generation of electricity and heat from the combustion of biomass (e.g. using a biomass integrated gasifier steam injection gas turbine system, BIG/STIG). The implementation of the Greenpeace FFES will necessitate the conquest of considerable technical, political and social barriers, points that should be emphasised by the student. Additionally, the Greenpeace FFES requires energy conservation policies if any shortfall in energy demand is to be met without fossil fuels and nuclear energy.

Answers to short questions

1 Atmospheric fluidised-bed combustion is a new technique used in coal-fired power stations to reduce noxious gaseous emissions. It involves passing the gases through a bed of coal and limestone that becomes supported (fluidised) by the upward flow of the gas in a fluidised-bed combustion furnace.

2 Renewable energy resources include hydroelectricity; wind power; tidal energy; wave power; solar energy; geothermal energy; and biomass energy.

3 Liquefied petroleum gas comprises mainly propane, butane and isobutane, but may contain unsaturated C_3 and C_4 hydrocarbons. Its main use is as a fuel, but it also has widespread use as an aerosol propellant, and as a chemical feedstock.

4 Attempts to reduce fuel consumption to improve air quality include a reduction in the emissions of greenhouse gases caused by the combustion of fossil fuels. This involves proposals for the introduction of a carbon tax, increased use of energy-efficient technologies, energy conservation, and a greater reliance on renewable energy resources.

5 Oilseed rape is used to make rape methyl ester (RME), which is used as a motor transport fuel, raw or blended, to power diesel engines.

6 The reserve lifetime for world coal is approximately 200 to 400 years and for world oil is about 56 years.

7 The main pollutants that result from motor vehicles include carbon dioxide; indirectly tropospheric ozone; carbon monoxide; nitrogen oxides; hydrocarbon compounds; chlorofluorocarbons; diesel particulates; lead; lead scavengers; aldehydes; benzene; non-diesel organics; asbestos; metals.

8 The ALTENER plan (= ALTernative ENERgy programme) is an EU five-year (1993–7) research and development plan to promote renewable energy sources within the community. Under the ALTENER plan, the EU has set targets for energy production from renewable energy sources for the year 2005.

9 With the construction of large dams, environmental problems include: the loss of vegetation and agricultural land; the displacement of the population; and changes in river courses that affect the animal and plant life. The reservoir may also create new habitats for parasites that carry diseases such as malaria and schistosomiasis. In addition, reservoirs may change ground-water tables, which may help to initiate earthquakes.

10 Nuclear fission involves the subdivision of a heavy atomic nucleus, for example uranium or plutonium, into two fragments of roughly equal mass, to release large amounts of energy. Nuclear reactions between light elements to form heavier ones to release large amounts of energy are known as nuclear fusion.

Additional notes on national energy resources and use

Students should be encouraged to research energy production and consumption patterns in a range of countries from the developed world to the developing world. They should research various access routes to such data and not rely solely on textbooks.

As an example, for Australian energy use and the value of its minerals and energy exports, students could contact the Australian Bureau of Agriculture and Resource Economics and the Australian Bureau of Statistics (Resources & Energy Group). For 1994–5, energy use in Australia totalled an estimated 4,285 PJ, broken down as follows: crude oil (36.4 per cent), black coal (28.4 per cent), brown coal (11.1 per cent), natural gas (18.1 per cent) and renewables (6 per cent). For the same period, 1994–5, the total value of Australia's minerals and energy exports is put at $30.3 billion, broken down as follows: coal ($6.9 billion), petroleum and gas ($3.8 billion), gold ($4.7 billion), bauxite, alumina and aluminium ($4.5 billion), iron ore and steel ($4.2 billion), and other ($6.2 billion). In 1994, Australia was the world's largest exporter of black coal, as well as iron ore, alumina, lead, diamonds and mineral sands products.

Students should compare the above figures for an economy in a developed country that has adopted a non-nuclear energy policy with say Japan or France, where the nuclear component of energy production and consumption is significant. From such analyses, the lecturer should encourage students to discuss how nations may have evolved different energy strategies and whether or not it is practicable and reasonable to expect any global harmonisation of energy policy at present or in the future. It is not important that a group of students reaches a consensus on such issues, but rather that they appreciate the complexity of energy strategies and that different countries have varying mixes of natural resources and differing access to renewables, etc.

Aims

- To introduce the concepts of hazard and risk analysis.

- To examine the causes and effects of natural hazards.

- To explore the various ways in which hazards can be mitigated.

- To assess if the magnitude and frequencies of natural hazards are increasing due to human activity and climate change.

Key point summary

- **Natural hazards** are typically unpredictable, and may cause death, injury, and destruction of and damage to agricultural land, buildings and communities. The effects of a natural disaster on a community depend upon factors such as the magnitude and extent of the disaster, how prepared the affected population is, and their economic resources to mitigate a potential disaster and/or clean up afterwards.

- The assessment of the magnitude and frequency of various natural disasters forms part of any risk assessment, and corresponding insurance.

- **Geological hazards** result from Earth's internal (tectonic) processes and Earth's surface (geomorphic) processes.

- **Earthquakes** and **volcanic eruptions** are a consequence of tectonic processes and usually occur along plate boundaries.

- Earthquake damage may be local and related to building collapse, fires, landsliding and subsidence, tsunamis, flooding, the release of poisonous gases, and associated hazards such as contaminated or depleted water sources, disease, famine, injury and death.

- Volcanic activity may result in regional and global climate changes. Earth scientists are developing more reliable means of predicting earthquakes and volcanic eruptions.

- **Geomorphologic hazards**, commonly included with geological hazards, include **landsliding**; **river flooding**; **glacial hazards**; **soil erosion**; and **asteroid and comet impacts**.

- The environmental effects depend upon their magnitude and frequency, which is a function of climate, geology, vegetation and human activity.

- Mitigating the effects depends upon a knowledge of the dynamics of Earth's surface processes, the accurate identification of high-risk zones, improved land-use practices, and the implementation of protective measures to reduce the effects.

- **Meteorological hazards** are driven by the Sun's energy and are controlled by atmosphere–ocean systems. The magnitude and patterns of weather systems control the severity of these hazards as well as the disaster preparedness of the threatened population.

- Meteorological hazards include **tornadoes, tropical cyclones, flooding** by heavy rainfall and storm surges, **disease** associated with contaminated water sources, **thunder and lightning** damage, **hail storms** and **droughts**.

- **Biological hazards** include **pests** and **environmental diseases**.

- Pests cause great destruction to agriculture and lead to the spread of disease. Such hazards can be controlled by physical, chemical and biological means, but most effectively by integrating all these approaches.

- Environmental diseases are one of the greatest hazards to health and can lead to millions of deaths each year. Amongst the most serious diseases to affect humans are **malaria, yellow fever, typhoid,** sleeping sickness and **schistosomiasis**. **Haem-orrhagic fevers**, such as **Ebola** and **Lassa**, are now being considered an increasing threat as human populations expand into marginal areas and cause ecological disruptions.

- Most diseases can be tackled successfully with good preventive education, early inoculation against disease, improved hygiene, and various artificial controls on the spread of the transmitting agent, for example spraying crops, etc. Prompt medical attention can help to cure individuals of many diseases.

Main learning hurdles

Definition of risk and hazard

Risk and hazard are commonly confused. The instructor should define these terms clearly so that the students can discuss the issues in this chapter precisely.

Tectonic hazards

The instructor should revise the section on plate tectonics in Chapter 1 to allow students to understand fully all the tectonic processes that are discussed in this chapter.

Meteorological hazards

The instructor should revise the sections on climate in Chapters 1 and 3 to allow students to understand fully all the meteorological processes that are discussed in this chapter.

Rates of natural processes

Students should be encouraged to appreciate the rates, magnitudes and frequencies of natural hazards. As an example the following list summarises typical orders of magnitude for the rates of movement of a range of natural slope-related instabilities (after Finlayson and Statham 1980):

Rockfalls	10^4 to 10^2 cm s^{-1}
Debris avalanches	10^4 to 10^2 cm s^{-1}
Air-supported mass flows	10^4 to 10^2 cm s^{-1}
Mudflows and debris flows	10^1 to 10^{-4} cm s^{-1}
Landslides and slumps	10^{-1} to 10^{-5} cm s^{-1}
Solifluction	10^{-5} to 10^{-7} cm s^{-1}
Creep	10^{-7} to 10^{-9} cm s^{-1}
Expansive soils	10^{-7} to 10^{-9} cm s^{-1}

Key terms

Accretionary prism; convergent plate boundary; debris flow; dilatancy-diffusion model; drought; dust bowl; Ebola; flash flooding; flowslide; glacial flood; haemorrhagic fever; heat engine; hot spot; hurricane; ice fall; ice shelf; iceberg; jökulhlaup; lahar; lake-water overturn; landslide; magma; malaria; mass movement; neoplastic disease; pest; plume; recurrence interval; retention pond; schistosomiasis (bilharzia); sensitive clay; sleeping sickness (African trypanosomiasis); snowstorm; spreading ridge; strain gauge; stream gauge; storm sewer; surging glacier; thunderstorm; tornado; tropical cyclone; typhoid; typhoon; volcano; yellow fever.

Issues for group discussion

Discuss the concept of safety.
The discussion should focus on how people perceive different types of hazard. This is dependent on the physical, social, economic and political environments in which they live and the discussion should consider these variables.

Discuss the view that natural hazards have increased during recent years.
Students should discuss the views that the apparent increase in hazards may be the result of increased media coverage, increased population growth and urbanisation, and global warming. Students should expand the latter to discuss the likely consequences of global warming and the evidence for change.

Discuss the role of the environmental scientist in mitigating natural hazards.
The discussion should focus on the way that environmental scientists monitor, measure and model natural systems and how these studies can aid in the reduction of hazards. The discussion should use examples to illustrate hazard reduction methods.

Selected readings

Begley, S. 1995. Lessons of Kobe. *Newsweek*, 125 (5), 16–27.
An interesting article that compares the effects of the 1995 Kobe earthquake with the 1994 Northridge earthquake. It describes various ways that earthquake damage may be mitigated.

Chapman, C.R. and Morrison, D. 1994. Impacts on the Earth by asteroids and comets: assessing the hazard. *Nature*, 367, 33–40.
This paper discusses the implications and assesses the hazard from asteroid and comet impacts.

Davies-Jones, R. 1995. Tornadoes. *Scientific American*, August, 48–57.
This review article discusses the dynamics of tornadoes and a current research project in the USA to help to understand and mitigate the hazard.

Dolan, J.F., Sieh, K., Rockwell, T.K., Yeats, R.S., Shaw, J., Suppe, J., Huftile, G.J. and Gath, E.M. 1995. Prospects for larger or more frequent earthquakes in the Los Angeles Metropolitan region. *Science*, 267, 199–205.
This research paper discusses the probability of future earthquakes in California.

Gregory, K. and Rowlands, H. 1990. Have global hazards increased? *Geographical Review*, 4, 35–8.
This is a useful article evaluating the possibility that global hazards have increased due to human activity or media coverage.

Grove, J.M. 1987. Glacier fluctuations and hazards. *The Geographical Journal*, 153 (3), 351–69.
This comprehensive paper describes most of the possible hazards associated with the glacial environment.

Knox, J.C. 1993. Large increases in flood magnitude in response to modest changes in climate. *Nature*, 361, 430–2.
This useful research paper presents sedimentological evidence that shows that during the past small changes in the climate of the Mississippi basin have resulted in quite large changes in the magnitude and frequency of flood events.

Le Guenno, B. 1995. Emerging viruses. *Scientific American*, October, 56–64.
This is a good review article discussing the nature and causes of haemorrhagic fevers.

Macilwain, C. 1994. Nature, not levees, blamed for flood. *Nature*, 369, 348.
This paper discusses the relative causes of the Mississippi floods.

Myers, M.F. and White, G.F. 1993. The challenge of the Mississippi floods. *Environment*, 38 (10), 5–9.
This comprehensive review outlines the history and nature of river management within the Mississippi basin.

Pendick, D. 1995. And here is the eruption forecast . . . . *New Scientist*, 7 January, 26–9.
This article discusses recent developments in analysing volcanic gases as a means to aid eruption forecasting.

Sarre, P. 1978. The diffusion of Dutch elm disease. *Area*, 10 (2), 81–5.
This research paper discusses the spread of Dutch elm disease throughout Britain.

Smith, K. 1993. Riverine flood hazards. *Geography*, 339, 78 (2), 182–5.
This paper is a simplified review of river flooding.

Spanier, E. and Galil, B.S. 1991. Lessepsian migration: a continuous biogeographical process. *Endeavour*, 15, 102–6.
This paper discusses the mixing of biogeographical provinces as a result of the construction of the Suez Canal, in the light of a plague of jellyfish.

Vaughan, D. 1993. Chasing the rogue icebergs. *New Scientist*, 9 January, 24–7.
This is an interesting article on monitoring iceberg hazards.

Wesson, R.L. and Wallace, R.E. 1985. Predicting the next great earthquake in California. *Scientific American*, 252 (2), 35–43.
Although this is a relatively old paper, it is a very interesting discussion of the use of palaeoseismic data in calculating recurrence intervals.

Textbooks

Blaikie, P., Cannon, T., Davis, I. and Wisner, B. 1994. *At Risk*. London: Routledge.
This book assesses and defines vulnerability, examining famines and droughts, biological hazards, floods, coastal storms, earthquakes, volcanoes and landslides. It draws together practical and policy

conclusions with a view to disaster reduction and the promotion of a safer environment. This is an essential text for all concerned with hazard mitigation and/or environmental impact assessment.

Bryant, E.A. 1991. *Natural Hazards*. Cambridge: Cambridge University Press.
An inter-disciplinary treatment of a variety of natural hazards that include oceanographic, climatological, geological and geomorphological hazards. This book is suitable as an introductory undergraduate text on natural hazards.

Chester, D. 1993. *Volcanoes and Society*. London: Edward Arnold.
An informative text on volcanic activity examined from geological, and socio-economic and political perspectives. This is well illustrated with many interesting examples of the processes and effects of vulcanicity worldwide. It is an ideal text for university and college students with Earth science and social science backgrounds wishing to pursue the issues associated with environmental risk assessment and natural hazards.

Hewitt, K. 1997. *Regions of Risk*. Harlow: Longman.
This book examines the various aspects of hazards, human vulnerability and disaster. It includes a variety of case studies on natural and technological hazards, as well as an examination of social violence. The book emphasises the cultural and social aspects of hazard assessment and response, as well as examining the cross-cultural differences and international scope of risk and disaster preparedness and response. This is a particularly useful text for geography students.

Smith, K. 1992. *Environmental Hazards: Assessing Risk and Reducing Disaster*. London: Routledge.
A comprehensive book covering most of the major environmental hazards, including seismic, mass-movement, atmospheric, hydrological and technological hazards. It integrates both the Earth and social sciences and is suitable for undergraduates from a variety of backgrounds, as well as being an important reference source for instructors and researchers.

Essay questions

1 'The magnitude and frequency of meteorological natural hazards have increased during the last few decades.' Discuss.

2 Do you believe that the developed nations provide sufficient support to meet the effects of natural hazards in the poorer developing nations?

3 Assess the regional and global environmental effects of tectonic geological hazards caused by earthquakes and volcanic activity.

4 What constitutes a natural disaster?

5 Discuss the various methods of combating the spread of diseases.

6 Evaluate the threats from diseases that are the result of environmental change.

7 Assess the various problems associated with the disruption of ecological systems.

8 'The most effective way to reduce the effects of earthquakes is through improved building design.' Discuss.

9 Discuss flood hazard reduction methods.

10 'Haemorrhagic fevers are one of the greatest threats to the existence of humankind.' Discuss.

11 Describe the characteristics of the recovery process after a disaster.

12 Assess the various methods for reducing the impact from EITHER flooding OR snowstorms.

Multiple-choice questions

Choose the best answer for each of the following questions.

1 Jökulhlaups are a type of:
(a) tropical cyclone
(b) glacial flood
(c) tidal wave
(d) volcanic mudflow

2 The best way to mitigate the effects of earthquakes is to:

(a) calculate recurrence intervals
(b) improve building design
(c) pump water into fault lines
(d) improve microseismic networks

3 Volcanic debris flows and mudslides are commonly called:
(a) lahars
(b) flowslides
(c) jökulhlaups
(d) pyroclastic

4 The 'factor of safety' is calculated to help to quantify:
(a) the stability of a slope
(b) seismic hazards
(c) the probability of a volcanic eruption
(d) meteorological hazards

5 Which of the following mass movements is the most rapid?
(a) flowslides
(b) creep
(c) debris flows
(d) mudslides

6 The strength of a tornado is compared using the:
(a) Mercalli scale of intensity
(b) Fujita intensity scale
(c) Richter scale
(d) Nazaki intensity scale

7 River discharge is measured using:
(a) stream gauges
(b) dischargemeters
(c) Pooh sticks
(d) stream-meters

8 A tropical storm is described as a tropical cyclone when the sustained wind speed exceeds:
(a) 33 km hr^{-1}
(b) 61 km hr^{-1}
(c) 91 km hr^{-1}
(d) 119 km hr^{-1}

9 The development of a hurricane can be explained using a model called the:
(a) tropical forcing effect
(b) heat engine
(c) Coriolis force
(d) thermohaline force

10 African trypanosomiasis is also known as:
   (a) sleeping sickness
   (b) Ebola
   (c) yellow fever
   (d) typhoid fever

11 Schistosomiasis (bilharzia) is spread by the fluke whose host is a:
   (a) mosquito
   (b) tsetse fly
   (c) snail
   (d) rodent

12 Haemorrhagic fevers are caused by:
   (a) viruses
   (b) bacteria
   (c) parasites
   (d) a, b and c

13 Which of the following is not a haemorrhagic fever?
   (a) Rift valley fever
   (b) Dengue fever
   (c) Gargal fever
   (d) Marburg fever

14 Tornadoes can be tracked using:
   (a) Doppler radar
   (b) air balloons
   (c) meteorological satellites
   (d) tornado gauges

15 Surging glaciers move at velocities in the order of several:
   (a) centimetres per day
   (b) metres per day
   (c) tens of metres per day
   (d) hundreds of metres per day

16 The probability of a large asteroid or comet (approximately 2 km diameter) colliding with the Earth within the next century is probably about:
   (a) 1 in 10,000
   (b) 1 in 100,000
   (c) 1 in 1,000,000
   (d) 1 in 10,000,000

17 Dutch elm disease is spread by a:
   (a) virus
   (b) bacterium
   (c) fungus
   (d) person from The Netherlands

18 The San Andreas Fault occurs along a:
   (a) divergent plate boundary
   (b) subduction margin
   (c) transform margin
   (d) convergent plate boundary

19 Icebergs are produced when:
   (a) sea-ice melts and breaks up
   (b) glaciers enter the sea and break up
   (c) the sea freezes in high latitudes
   (d) a, b and c

20 Sensitive clays are sediments that:
   (a) are emotionally stressed
   (b) deform and fail under stress
   (c) are very resistant to failure
   (d) have been deformed and are resistant to failure

21 Urbanisation affects the hydrology of a region by:
   (a) increasing the frequency of flooding
   (b) increasing the magnitude of flooding
   (c) reducing the lag time between maximum rainfall and peak flood event
   (d) a, b and c

22 The human-induced earthquakes experienced in Denver, Colorado, during the period 1962–5 were the result of:
   (a) underground nuclear explosions at the Rocky Mountain Arsenal
   (b) conventional arms testing at the Rocky Mountain Arsenal
   (c) underground pumping of waste at the Rocky Mountain Arsenal
   (d) reservoir construction

23 The major cause of damage and fatalities during the Gansu earthquake in central China in 1920 was:
   (a) the spread of fires
   (b) the collapse of buildings
   (c) landslides
   (d) floods

24 Most flash flooding is the result of:
   (a) dam burst
   (b) glacier floods
   (c) increased urbanisation
   (d) climatic conditions

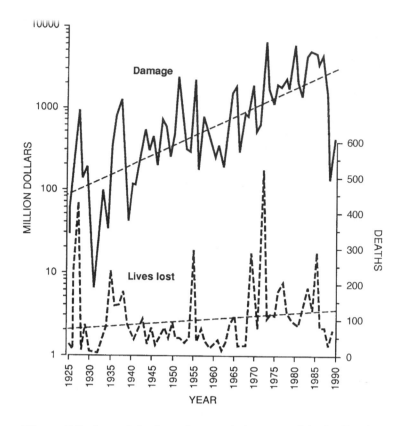

Figure 8.8 *Annual deaths and economic losses caused by flooding in the United States, for the years 1925–89. Damages are in US$ millions adjusted to 1990 values. Redrawn after Smith (1992).*

25 Neoplastic diseases are medical conditions that are probably induced by:
  (a) pests
  (b) plastics
  (c) radioactive water
  (d) viruses

Figure questions

1 Figure 8.8 shows the annual deaths and economic losses caused by flooding in the USA, for the period 1925–89. Answer the following questions.
  (a) Describe the characteristics of the data presented.
  (b) Suggest reasons for the increase in deaths and losses since 1925.
  (c) List the basic methods that may be used to help to mitigate the effects of flooding.

2 Figure 8.9 illustrates the principal cause of floods and the factors contributing to intensifying conditions. Explain how the intensifying factors might operate.

9 Figure 8.12 shows the losses of (A) property, and (B) life in the continental United States due to tropical cyclones for the periods 1915–89 and 1900–89, respectively. Answer the following questions.
  (a) Suggest reasons for the inverse relationship between damage and deaths.
  (b) What would you expect graphs to be like for the losses of property and life for tropical cyclones in Bangladesh?
  (c) Suggest means to reduce the effects of tropical cyclones.

Short questions

1 What are the Fujita and Mercalli intensity scales?

2 Describe the characteristics of flash flooding.

3 What is Dutch elm disease?

4 Describe the symptoms of haemorrhagic fevers.

5 List the main types of geomorphological hazard.

6 What is a seismic gap?

7 Describe the different types of mass movement.

8 List the types of hazard associated with volcanoes.

9 How can building construction be improved to reduce the effects of an earthquake?

10 Describe the main effects of a tropical cyclone.

Answers to multiple-choice questions

1 b; 2 b; 3 a; 4 a; 5 a; 6 b; 7 a; 8 d; 9 b; 10 a; 11 c; 12 a; 13 c; 14 a; 15 b; 16 a; 17 c; 18 c; 19 b; 20 b; 21 d; 22 c; 23 c; 24 d; 25 c.

Answers to figure questions

1 (a) The running means for these data show that the number of deaths has increased gradually, but the amount of damage has increased exponentially throughout this century. The data also show exceptionally large loss of life during particular years, for example 1972. This is probably because of an exceptionally large flood event. (b) The increase in death

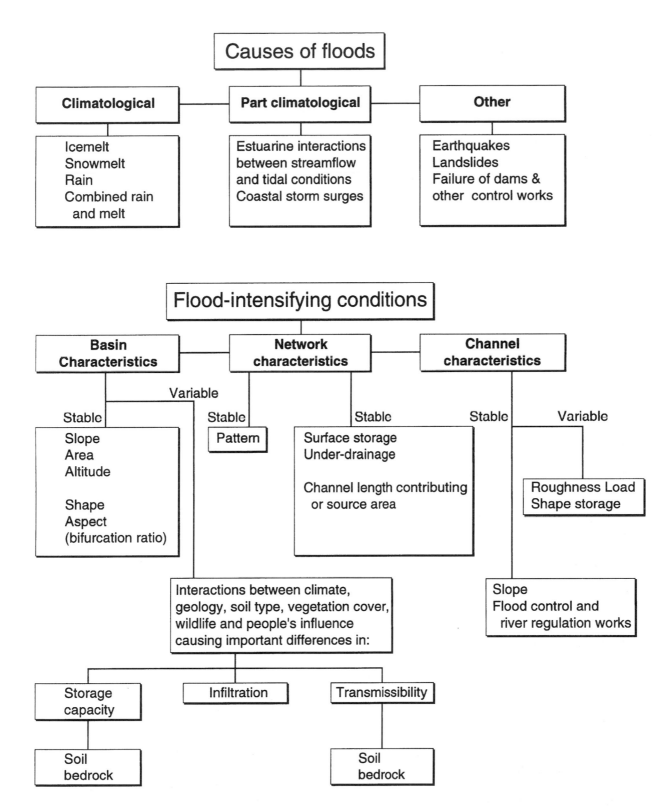

Figure 8.9 *The causes of floods and flood-intensifying conditions. Redrawn after Ward (1978).*

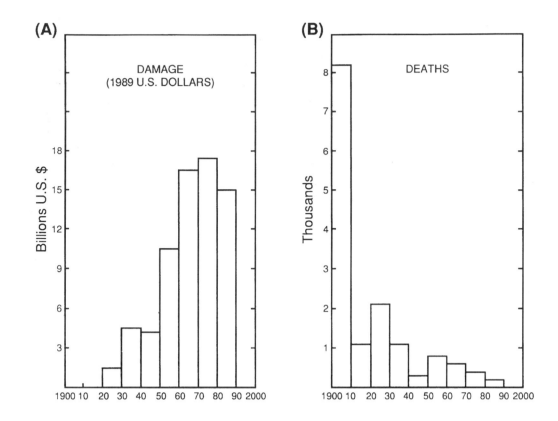

Figure 8.12 *Losses of (A) property and (B) life in the continental United States due to tropical cyclones for the periods 1915–89 and 1900–89, respectively. Redrawn after Gross (1991).*

and losses may be due to an increase in the incidence and magnitude of flood events over this time period, but this is unlikely and it is probably a result of increased population growth and wealth, so that a flood today causes more damage because there is more to damage than in the past. The exponential growth in damage, as compared to linear growth in loss of life, suggests that warning and evacuation mechanisms have improved faster than the rate of population growth. (c) The basic methods to reduce flooding involve river basin management schemes, which include monitoring weather and flow conditions, and issuing warnings of flood events; flood prevention measure, such as channelisation schemes and improved levee construction; and controlling development in flood-prone areas.

2 See Fig. 8.10.

3 (a) The inverse relationship between damage and death is the result of an increase in wealth, such that more is destroyed during a hurricane today and the decrease in death is a result of improved hurricane

warning and protection schemes. (b) The graphs for damage and loss of life due to hurricanes in Bangladesh would both show a positive increase. The damage in terms of loss of wealth would be significantly smaller that for the USA, but the loss of life would be very high because of the dramatic increase in population in Bangladesh since the 1950s and the difficulties of implementing an effective hurricane warning system and prevention programme in poor developing countries. (c) The main means to reduce the effect of tropical cyclones is implementing a protective warning system and providing safe places to evacuate the people. To reduce the risk of disease immediately after the hurricane, relief workers are needed to provide food, fresh water and medical supplies. The effects of flooding can be reduced by building sea walls and improving river protection schemes. Where possible or realistic, future developments might be encouraged to be sited away from low-lying areas that are prone to flooding. Little can be done to reduce the effects of the wind, except that early warnings might help people to board up windows and bring loose objects indoors.

Answers to short questions

1 The Fujita intensity scale describes the damage associated with a tornado. It relates to the velocity of the rotating spiral of wind. The Mercalli intensity scale describes the damage associated with earthquakes.

2 Flash flooding is commonly associated with ephemeral streams in arid and semi-arid environments. An almost instantaneous rise in discharge that progresses downstream as waves, for example, or as bores (solitary waves) characterise such a flood.

3 Dutch elm disease is a fungoid killer of elm trees. It was introduced into Europe from Asia during the Second World War and was first found in The Netherlands.

4 Haemorrhagic fevers are characterised by an illness in which the patient initially develops a fever. This precedes a period in which the patient deteriorates and superficial bleeding develops, where blood seeps from vessels under the skin and bruises appear. Other cardiovascular, digestive, renal and neurological complications follow. In the most serious cases, the patient dies from massive haemorrhages or sometimes multiple organ failure.

5 The main types of geomorphological hazard include: landsliding and subsidence; river flooding; glacial hazards (floods, advances/retreats, ice falls); soil erosion; and coastal erosion.

6 A seismic gap is a zone within a tectonically active region that has not experienced large earthquakes during historic time or the recent past. Stress builds up and may release catastrophically within these zones. These zones, therefore, constitute the greatest tectonic hazard.

7 Mass movements may transport a variety of different materials, ranging from large blocks of rock and/or snow and ice to sediment and soils. The movement involves a variety of mechanisms, ranging from free fall and avalanche to fluidisation, sliding, slurring and creeping. Mass movement classifications include both the material and the nature of the movement. Most mass movements are complex, involving several types of mechanism and material.

8 The types of hazards associated with volcanoes include: advancing lavas; fires; falling pyroclastic debris; poisonous gases; lahars and landslides; earthquakes; tsunamis; and climatic changes associated with aerosols and gases.

9 To reduce the effects of an earthquake, improvements to building construction fall into four main categories. First, the basement and foundations can be mounted on rubber to permit the ground to move below and so help to isolate the building from the tremors. Second, a non-eccentric layout of buildings can be used, since irregular internal and external layout can result in complex torsional and swaying movements, which cause greater damage. Third, the buildings can be made flexible to allow the structure to flex and deform rather than break and collapse. Finally, the ground floor can be strengthened to help to support the overlying structure, rather than the more typical structurally weaker ground-floor design with large open spaces, many doorways, vestibules, and large window areas.

10 A tropical cyclone affects a region in three main ways. First, the very fast winds cause destruction to buildings, vegetation and loose objects. Second, heavy rainfall causes extensive flooding, both along coasts and inland. Third the sea surge results in flooding along coastal regions.

Additional references

Dade, W.B. and Huppert, H.E. 1996. Emplacement of the Taupo ignimbrite by a dilute turbulent flow. *Nature*, 381, 509–10.

Dade and Huppert describe the 1,800-year BP Taupo ignimbrite, North Island, New Zealand, which travelled outward from the volcanic eruption at c. 200 m s^{-1}, as an avalanche-like flow with a volumetric solids concentration of greater than 30 per cent. This was one of the largest volcanic eruptions during the past 10,000 years.

Laird, K.R., Fritz, S.C., Maasch, K.A. and Cumming, B.F. 1996. Greater drought intensity and frequency before AD 1200 in the Northern Great Plains, USA. *Nature*, 384, 552–4.

This paper presents a reconstruction of the intensity and frequency of droughts over the past 2,300 years for the Northern Great Plains in North America. One such extreme drought occurred in the 1930s in the so-called 'Dust Bowl'. The sub-decadal resolution, based on diatom-inferred salinity changes in Lake Moon within the topographically closed-basin, eastern North Dakota, shows a significant change at about AD 1200. Extreme droughts, of greater intensity than that of the 1930s, appear to have been more frequent prior to AD 1200, the more recent past being dominated by wetter/cooler conditions. In the time interval studied, the most extreme droughts occurred between AD 200–370, AD 700–850 and AD 1000–1200. The change coincides with the termination of the 'Medieval Warm Period' and the onset of the 'Little Ice Age' in North America, between AD 1300–1850. The authors propose that the climatic shift at *c.* AD 1200 was a consequence of changing atmospheric circulation patterns that produce extreme droughts.

Rose, W.I., Delene, D.J., Schneider, D.J., Bluth, G.J.S., Krueger, A.J., Sprod, I., McKee, C., Davies, H.L. and Ernst, G.G.J. 1995. Ice in the 1994 Rabaul eruption cloud: implications for volcano hazard and atmospheric effects. *Nature*, 375, 477–9.

Satake, K., Shimazaki, K., Tsuji, Y. and Ueda, K. 1996. Time and size of a giant earthquake in Cascadia inferred from Japanese tsunami records of January 1700. *Nature*, 379, 246–9.

This paper describes an inferred earthquake in the Cascadia subduction zone, west coast USA, of magnitude 9 on 26 January 1700, probably occurring at *c.* 2100 hours, producing a tsunami along the Japanese coast.

Assignment

Students should be encouraged to undertake a literature search and produce an essay, dissertation and/or class/group presentation on a current or very recent natural hazard. They should find out about the origin of a particular natural hazard, the risks posed and potential consequences of the hazard being realised. Where possible, students should find an Environmental Impact Assessment of the selected hazard and, where appropriate, a hazard map.

For an exercise of this type, students will have to use information from the media (newspapers, journals, radio and television programmes, and the World Wide Web pages). The purpose of such an exercise should include an appreciation of the direct relevance of studying natural hazards, an understanding of risk assessment and risk management.

An example of a current natural hazard (1996) is the eruption of the Soufriere Hills volcano on the island of Montserrat in the West Indies. The eruption began on 18th July 1995. The volcano produces viscous magmas which construct tall, steep and unstable domes above the vent. These domes collapse to form large hot avalanches that can travel at speeds of up to 100 km hr^{-1}, and it is these flows that constitute the greatest hazard to those living in the vicinity of the volcano.

In the case of Montserrat (or a comparable study), students can compile an inventory of the volcanic activity, find a hazard map, and evaluate the methodological approaches that were used in the risk assessment. Students should appreciate the uncertainties associated with the risk assessment.

An alternative suggestion is for students to investigate the October/November 1996 sub-glacial volcanic eruption of Grimsvötn beneath the Vatnajökull ice-cap, Iceland. The eruption was important for contributing to our understanding of how sub-glacial explosive volcanic eruptions cause the melting of very large volumes of ice which accumulate as subglacial fresh-water lakes that can suddenly drain – in this particular case over about 36 hours – resulting in a catastrophic jökulhlaup event which swept away roads and bridges. The bridge supports buckled and failed under the impact of large icebergs.

Aims

- To examine the main environments on the Earth's surface and the processes that operate within them.

- To assess the impact of human activity on the Earth's natural environments throughout time.

- To evaluate the environmental impact of human activity and methods to reduce the impact.

Key point summary

- Rapid population growth has resulted in increased demands upon the Earth's resources, which has led to accelerated **environmental degradation**, and precipitated potentially serious global climate change.

- The human impact on land has been enormous. As land use changes, natural vegetation is cleared for agricultural use, settlements and urbanisation increase, reservoirs are created, minerals are extracted, and more land is developed for recreation purposes.

- Acute concern is now widely expressed over the **deforestation** of **boreal** and **tropical forests**, the **degradation** of **grasslands** and **wetlands**, and **desertification**. Such destruction of natural ecosystems leads to a reduction in biodiversity and impoverishment of soils.

- Attempts to counter the deleterious effects of land misuse in some areas include the introduction of exotic plants and animals, and the monitoring of indigenous fauna and flora.

- Human impact on **soils** has caused considerable damage. This is commonly because of poor agricultural practices, excessive water extraction, poor irrigation methods (for example, leading to **salinisation**), defoliation (particularly resulting in **laterisation**), and compaction by heavy vehicles

and animals. The cumulative effects of these can be disastrous for countries whose economies are heavily dependent on agriculture.

- The amelioration of these poor land-use practices and improved soil quality require an understanding of the chemistry of soils and nutrient supply cycles.

- Wetlands contribute almost a quarter of the world's primary productivity and are essentially the interface between terrestrial and aquatic environments. Only recently have wetlands been considered valuable and attempts to reduce their destruction been implemented.

- Human impact on the oceans and seas results from pollution by dumping and accidents, over-fishing, mineral extraction (for example, phosphates) and the removal of rare and important marine life such as corals.

- The seas are an available resource that requires more careful research in order to avoid irreversible damage to their ecosystems, which could have a knock-on effect to the atmosphere and, ultimately, terrestrial life.

- **Environmental risk management (ERM)** involves the evaluation of the hazards and impacts on the environment. **Environmental impact assessments (EIAs)**, **environmental audits** and **legislation** involve a multi-disciplinary approach.

- The exploitation of Earth's resources inevitably produces waste, some of which may be hazardous/toxic (contaminants).

- Until the past few decades, much of this waste has been disposed of without any real concern for the damage to ecosystems, and frequently under the auspices of 'not in my back yard'. Clean-up technologies are more readily available and preventive measures are being instigated in many countries. Many nations and international organisations are adopting a philosophy of the 'polluter pays'. Responsibility for cleaning up contaminated land has led to the introduction of legislation in countries such as the USA and throughout Europe. New and forthcoming legislation aims to identify the polluters and arrange for appropriate levels of compensation to injured parties. However, because such laws are in their infancy, there are many teething problems, exemplified by the US Superfund.

Main learning hurdles

Soil processes

Students may be confused by pedological nomenclature, particularly the classification of soils. The instructor should, therefore, try to simplify the discussion on soils by referring to the basic processes of illuviation and elluviation, cation exchange, leaching and particularly important processes such as salinisation and laterisation.

Nutrient cycling

Nutrient cycling within ecosystems is complex, involving soils, plants, the atmosphere and the hydrosphere. The instructor should revise the discussion on nutrient cycles that is presented in Chapter 1 and show how these relate to changes in the biosphere and soils that result in land degradation.

Erosion

This chapter assumes that the student has a basic knowledge of the main processes of erosion. The instructor should, however, be certain that the students understand the basics of erosional processes, which include abrasion, corrasion, corrosion, cavitation and attrition. The students should also appreciate the basic mechanisms that transport sediment and processes of deposition. For further explanation, the students are advised to see geomorphological textbooks such as Cooke and Doornkamp (1990).

Key terms

Badlands; black smoker; bromeliad; campo rupestre; cerrado; contaminant; contaminated land; deforestation; desertification; Dutch elm disease; erodibility; erosivity; evapotranspiration; land degradation; laterisation; longshore sediment drift; manganese nodule; nutrient cycling; rain splash erosion; salinisation; savannah; urbanisation; wetlands.

Issues for group discussion

Discuss the various data that are needed to produce a good environmental impact assessment.
This discussion should emphasise the importance of having scientific, social, economic and political information for a full environmental impact analysis. The students should discuss the detailed scientific information, for that is used in a GIA. This includes geological, pedological, hydrological, geomorphological, climatological and biological data.

Discuss which aspect of human activity is the most environmentally devastating.
The students should list the main threats, including the impact on the atmosphere, biosphere, hydrosphere and lithosphere. The discussion should centre around how these spheres are interlinked. The students should emphasise that it is not really possible to assess which activity has the greatest impact because of the complex interactions.

Discuss the various international strategies for reducing human impact on the Earth's surface.
The discussion should consider the various international conventions that help to reduce human impact on the Earth's surface, particularly the World's Conservation Strategy and Agenda 21. Students should discuss the effectiveness of the UNEP and the WCED.

Selected readings

Binns, T. 1990. Is desertification a myth? *Geography*, 75 (2), 106–13.
This is an interesting discussion of the evidence for desertification.

Burman, A. 1991. Saving Brazil's savannahs. *New Scientist*, 2 March, 30–4.
This is an interesting article on the destruction of Brazil's savannahs.

Hodges, C.A. 1995. Mineral resources, environmental issues and land use. *Science*, 268, 1305–12.
This article describes the world's mineral reserves and resources, highlighting that given that there is

currently no shortage of mineral resources, mining activities should be considered not only from their economic importance but also from their environmental impact.

Hughes, T.P. 1994. Catastrophes, phase shifts, and large-scale degradation of a Caribbean coral reef. *Science*, 265, 1547–51.
This research paper summarises the factors resulting in the degradation of coral reefs, illustrated using reefs in the Caribbean.

Hulme, M. and Kelly, P.M. 1993. Exploring the links between desertification and climate change. *Environment*, 35 (6), 4–11 and 39–46.
This paper summarises the causes of desertification and the associated problems.

Keller, E.A. 1976. Channelisation: environmental, geomorphic and engineering aspects. In: Coates, D.R. (eds), *Geomorphology and Engineering*, 115–40. Stroudsburg: Dowden, Hutchinson & Ross.
A good summary of the environmental problems associated with channelisation, and well illustrated.

Orians, G.H. 1995. Cumulative threats to the environment. *Environment*, 37 (7), 6–14 and 33–6.
This is a thoughtful paper examining the cumulative effects of environmental degradation. It is written in a very accessible manner and contains a useful reference list.

Mabogunje, A.L. 1995. The environmental challenges in Sub-Saharan Africa. *Environment*, 37 (4), 4–9 and 31–5.
This article provides a thorough examination of the problems of desertification and the consequent challenges that face the societies and economies of sub-Saharan Africa.

MacKenzie, D. 1995. The cod that disappeared. *New Scientist*, 16 September, 24–9.
This is a useful article on the depletion of fish stocks on the Grand Banks. It argues that current biological models for predicting the sizes of fish populations are inadequate.

Micklin, P.P. 1988. Desiccation of the Aral Sea. *Science*, 241, 1170–6.
This is a good summary of the causes and problems associated with the desiccation of the Aral Sea.

Milliman, J.D. *et al.* 1987 Man's influence on the erosion and transport of sediment by Asian rivers: The Yellow River example. *Journal of Geology*, 95, 751–62.
This rather technical paper discusses the effect of agricultural activity in the Yellow River basin on sedimentation rates.

Pearce, F. 1994. Rush for rock in the Highlands. *New Scientist*, 8 July, 11–12.
This interesting article examines the issues surrounding the construction of superquarries in Scotland to provide rock for aggregates and construction.

Pearce, F. 1993. When the tide comes in. *New Scientist*, 2 January, 22–7.
This article assesses the value of constructing coastal defences around the British coastline.

Textbooks

Angotti, T. 1993. *Metropolis 2000: Planning, Poverty and Politics*. London: Routledge.
This very readable book on urbanisation offers an analysis of metropolitan development and planning in all parts of the world, and under different economic and environmental conditions. The first four chapters examine metropolitan development in the United States, the former Soviet Union, and the 'dependent metropolis' of the developing world. The last three chapters consider the problems of urban planning theory and practice in the metropolis and its communities. Throughout, Angotti advances the principle of 'integrated diversity' and emphasises linked neighbourhood planning with a broader vision of a planned metropolis.

Cooke, R.U. and Doornkamp, J.C. 1990. *Geomorphology in Environmental Management*, second edition. Oxford: Oxford University Press.
A comprehensive text that highlights the importance of geomorphology in environmental management and risk assessment. This book is suitable for college and university undergraduate students at all levels, and provides a useful reference source for instructors and researchers. Topics include mass movement, catchment studies, erosion and weathering problems, neotectonics, aeolian environments, and glacial systems.

Eden, M.J. 1989. *Land Management in Amazonia*. London: Belhaven.
This book considers the tropical rainforest as a global resource and its vital importance in sustaining life on Earth. The book assesses, in terms of climate, geomorphology, hydrology, soils, ecology and diverse

histories, and current impact of human intervention, the competing needs of conservation and development in Amazonia. Case studies present examples of development and management schemes. For example, there are attempts to adapt resource-use systems of native peoples to encourage the more effective and less harmful exploitation of the rainforests. The book also addresses conservation issues, including the role of national parks and interpretative land management. A good supplementary book for students concerned with environmental risk assessment, particularly relating to the rainforests.

Eden, M.J. and Parry, J.T. (eds) 1996. *Land Degradation in the Tropics*. London: Cassell Academic.
This is a wide-ranging, coherent and scholarly account of land degradation in the tropics. It emphasises the integration of information and theory from both the environmental and social sciences. It also illustrates the application of scholarly analysis in actual policy formulation, planning and management. A wide variety of case studies illustrate the following headings: degradation of tropical forests; degradation in the drier tropics; degradation in tropical wetlands; and urban and industrial degradation in the tropics. This is a useful text for planners, managers and policy-makers, as well as students of environmental science.

Ellis, S. and Mellor, A. 1995. *Soils and Environment*. London: Routledge.
This book examines the ways in which soils are both influenced by, and themselves influence, the environment. In addition to describing the analysis of soil properties, soil processes and classification, it discusses soil–human interactions and examines the relation to land systems, environmental problems and management, soil surveys, and land evaluation. This is a useful text for students of all levels.

Goudie, A. 1993. *The Human Impact: On the Natural Environment*, fourth edition. Oxford: Blackwell Scientific.
A very useful undergraduate textbook that addresses the ways that human activity has changed, and is altering, the Earth's surface. The book is well illustrated and includes a comprehensive bibliography. Topics include desertification, deforestation, plant and animal invasions, marine pollution, climatic change, and environmental uncertainty.

Gradwohl, J. and Greenberg, R. 1988. *Saving Tropical Forests*. London: Earthscan.
This text provides case studies from throughout the world to show how the destruction of the tropical

forests might be slowed or even stopped, and how sustainable management could be achieved. It is a thought-provoking book for students, instructors and policy-makers.

Grainger, A. 1990. *The Threatening Desert: Controlling Desertification*. London: Earthscan.
This is an interesting book that describes the distribution and processes of desertification, and the successes and failures that have accompanied the various attempts to combat desertification as set out by the Plan of Action resulting from the 1977 Nairobi United Nations Conference on Desertification. The book argues for a new international plan of action to control the increasing threat to the natural environment posed by desertification.

Ives, J.D. and Messerli, B. 1989. *The Himalayan Dilemma*. London: Routledge.
This book addresses the complex dynamics and environmental systems in the Himalayas, and considers the problems of reconciling development and conservation. The book includes a look at the interaction between human activities and the natural environment. It is a useful book for advanced undergraduate courses, instructors and policy-makers.

Morgan, R.P.C. 1986. *Soil Erosion and Conservation*. London: Longman.
A useful textbook aimed at undergraduate and postgraduate students who are studying soil erosion and conservation as part of any Earth science or environmental science course. The book provides an introduction to the subject, including the magnitude, frequency, rates and mechanics of wind and water erosion; erosion hazard assessment; methods of measurement, modelling and monitoring; and strategies for erosion control and conservation practices.

Parnwell, M. and Bryant, R. (eds) 1996. *Environmental Change in South-East Asia*. London: Routledge.
This edited book explores the interaction of people, politics and ecology. It examines the nature of the environmental degradation that has resulted from the rapid economic growth in Southeast Asia and the dilemmas facing policy-makers as they seek to promote sustainable development.

Poore, D. 1989. *No Timber Without Trees: Sustainability in the Tropical Forest*. London: Earthscan.
Based on a study for the International Tropical Timber Organisation, this book reviews the extent to which natural forests are being sustainably

managed for timber production and how these practices could be improved. The book places timber production in the wider context of tropical rainforest conservation. It is illustrated with examples from Queensland, Africa, South America, the Caribbean and Asia. The book makes interesting and easy reading for students of environmental science.

Revkin, A. 1990. *The Burning Season: The Murder of Chico Mendes and the Fight for the Amazon Rain Forest*. London: Collins.
An inspiring, but also sad, book that emphasises the beauty of the tropical rainforests and the need for conservation in Amazonia. The book provides an ecological, historical and industrial outline of life in the forest, focusing on the life and environmental work of Chico Mendes, a rubber planter, who strove for sustainable development in the forest in which he lived. Mendes' success in reducing the exploitation of the forest by cattle ranchers cost him his life when he was murdered in 1988.

Smith, L.G. 1993. *Impact Assessment and Sustainable Resource Management*. Harlow: Longman.
This book provides an integrated approach to environmental planning, balancing academic and practical considerations. Various aspects of environmental planning include decision-making, dispute resolution, environmental law, public policy, administration, the nature of planning, impact assessment and methodology. This is a useful text for planners, managers and policy-makers.

Thomas, D.S.G. and Middleton, N.J. 1994. *Desertification: Exploding the Myth*. Chichester: Wiley.
A useful text that explores the origin of the 'desertification myth' and how it has spawned multi-million dollar initiatives to become regarded as a leading environmental issue. The book examines the political and institutional factors that created the myth, sustaining it and protecting it against scientific criticism.

Tivy, J. 1993. *Biogeography: A Study of Plants in the Ecosphere*, third edition. Harlow: Longman.
This is the third edition of the classic text on biogeography. It explores the variations in forms and functioning of the biosphere at both the regional and global scales. It highlights the interaction between the organic and inorganic components of the ecosphere. The book emphasises the importance of the plant biosphere as the primary biological product that forms the vital food link between organisms. It also emphasises the role of humans as the dominant ecological factor. It is an essential undergraduate textbook for those studying geography, biology, environmental studies, conservation and ecology.

Welford, R. 1995. *Environmental Strategy and Sustainable Development*. London: Routledge.
This is an interesting debate over environmental strategy in business, providing a radical business agenda for the future. It discusses important strategies such as environment management systems and environmental audits.

Williams, M.A.J. and Balling, R.C. 1995. *Interactions of Desertification and Climate*. London: Edward Arnold.
This book, commissioned by the UNEP and the WMO, examines the current knowledge of the interactions of desertification and climate in drylands. It provides a series of useful recommendations for future dryland management.

Essay questions

1 'Sub-Saharan Africa suffers from some serious environmental problems, including deforestation, soil erosion, desertification, wetland degradation, and insect infestations. Efforts to deal with these problems, however, have been handicapped by a real failure to understand their nature and possible remedies' (Mabogunje 1995, 4). Discuss.

2 Describe the environmental consequences of the exploitation of mineral resources and measures to reduce any damaging environmental impact.

3 Discuss the possible regional and global implications of the continued destruction of the rainforests.

4 It is difficult to retard development and enforce conservation in less developed counties when people are merely subsisting or dying of starvation. How may this dilemma be reconciled?

5 Describe the causes of soil degradation and the environmental consequences.

6 With reference to land contamination, discuss the 'polluter-pays' principle.

7 'The most widespread and important cumulative effect of human activities in terrestrial environments is landscape modification' (Orians 1995, 8). Discuss.

8 Evaluate the importance of wetlands as buffers to environmental change both regionally and globally.

9 Assess the role of urbanisation as a major agent in environmental degradation in coastal regions.

10 Describe how the introduction of exotic fauna and flora may lead to environmental stress in an ecological system and result in environmental degradation within such a region.

11 Outline the properties that determine the fertility of a soil and describe the ways in which human management modifies these properties.

12 'Soils and soil processes have had a profound impact on the development of agriculture, and civilisation, throughout the ages. Mistreatment of soils has, over the millennia to the present day, been responsible for horrific human tragedy' (Greenwood 1993). Discuss.

13 With reference to specific examples, examine the environmental consequences of deforestation.

14 'The management of environmental risk is significantly governed by the choice of response available.' Discuss.

15 'Soils can be viewed as an environmental product, moulded over time from whatever material was available, by climate, organisms and geomorphic processes' (Retallack 1990). Discuss.

Multiple-choice questions

Choose the best answer for each of the following questions.

1 An intensely dissected landscape produced by natural or human-influenced erosion is known as:
   (a) desert
   (b) drylands
   (c) badlands
   (d) scablands

2 The process that involves the enrichment of sesquioxides of aluminium and/or iron in a soil is known as:
   (a) podsolisation
   (b) salinisation
   (c) aluminisation
   (d) laterisation

3 Loess is a:
   (a) sediment rich in aluminium sesquioxide
   (b) sediment comprising dominantly silt
   (c) lake sediment
   (d) sediment comprising dominantly sand

4 Which of the following sediments is the most erodible?
   (a) river gravel
   (b) loess
   (c) glacial clays
   (d) river sands

5 Pedology is the study of:
   (a) soils
   (b) erosion
   (c) rivers
   (d) ground water

6 The Comprehensive Response, Compensation, and Liability Act (CERCLA) is also known as:
(a) Superfund
(b) EPA fund
(c) Remedial fund
(d) Megafund

7 Salinisation can be caused by:
(a) irrigation
(b) adding fertilisers to the soil
(c) adding lime to the soil
(d) a, b and c

8 Degradation of the Aral Sea is mainly the consequence of:
(a) climate change
(b) pollution
(c) irrigation
(d) over-fishing

9 Deforestation of tropical rainforests results in:
(a) laterisation
(b) increased lead poisoning
(c) disruption of the nutrient cycles
(d) a, b and c

10 Wetlands cover approximately:
(a) 6 per cent of the Earth's land surface
(b) 16 per cent of the Earth's land surface
(c) 36 per cent of the Earth's land surface
(d) 66 per cent of the Earth's land surface

11 The contribution by wetlands to the world's primary productivity is approximately:
(a) 14 per cent
(b) 24 per cent
(c) 34 per cent
(d) 44 per cent

12 The spread of Dutch elm disease across the British countryside was accelerated by:
(a) a hot summer
(b) the transport of diseased wood
(c) water pollution
(d) a, b and c

13 The opening of the Suez Canal in 1869 resulted in:
(a) increased pollution in the Mediterranean Sea
(b) the sudden mixing of former isolated biogeographical communities
(c) temperature changes in the Mediterranean Sea
(d) a, b and c

14 The degradation of permafrost results in:
(a) thermokarst
(b) tundra
(c) cryogenesis
(d) cryoturbation

15 The Resources, Conservation and Recovery Act (RCRA), passed in 1976, was ratified to help to:
(a) reduce pollution
(b) conserve biodiversity
(c) inhibit deforestation
(d) a, b and c

16 Human activity has been affecting landscapes since:
(a) Tertiary times
(b) Palaeolithic times
(c) Neolithic times
(d) the Industrial Revolution

17 Which of the following is not a savannah?
(a) cerrado
(b) caatinga
(c) campo rupestre
(d) catina

18 During the Last Glacial Maximum, the Sahara Desert:
(a) was less extensive
(b) was more extensive
(c) was as extensive as now
(d) did not exist

19 *Rhopilema nomadica* is a type of:
(a) rodent
(b) fish
(c) jellyfish
(d) plankton

20 Submarine hydrothermal chimneys comprising metal sulphides are called:
(a) atolls
(b) parasitic vents
(c) fumaroles
(d) black smokers

21 Present mineral reserves are:
(a) not adequate for current needs
(b) adequate for current needs
(c) more than adequate for current needs
(d) not possible to calculate

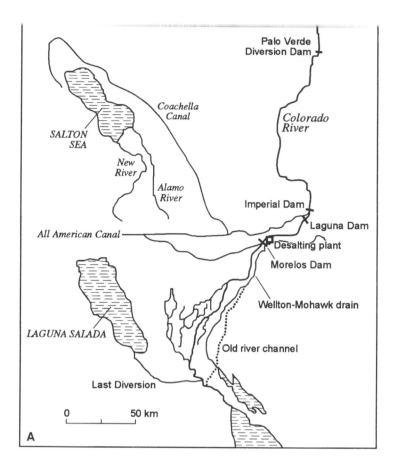

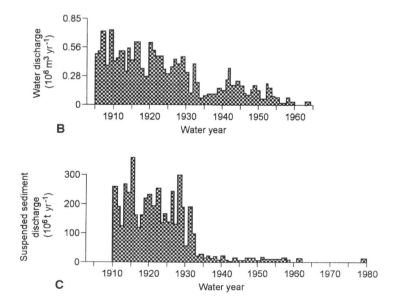

Figure 9.17 *(A) Dam constructions along the lower part of the Colorado River; and (B) discharge variations and (C) sediment yield variations throughout the twentieth century resulting from the damming. Redrawn after Graf (1987) in Petts (1994).*

22 The once rich fishing grounds in the North Atlantic off the coast of Newfoundland are called the:
  (a) Grand Banks
  (b) Great Banks
  (c) Great Grounds
  (d) Grand Shelf

23 The view that deforestation in the Himalayas has caused increased flooding in Bangladesh:
  (a) has been proven true
  (b) has been proven false
  (c) needs to be proved
  (d) is not worth proving

24 Urbanisation results in:
  (a) increased flooding
  (b) changes in local climate
  (c) increased sediment loads in streams
  (d) a, b and c

25 Deliberate burning to clear land may lead to:
  (a) improved soil quality
  (b) reduction of widespread fires
  (c) soil erosion, flooding and wind erosion
  (d) a, b and c

Figure questions

1 Figure 9.17 illustrates the effects on discharge and sediment yield due to dam construction along the lower part of the Colorado River. Answer the following questions.
  (a) Why does discharge decrease steadily from 1910 to 1960, whereas there is an abrupt decrease in sediment load after the early 1930s?
  (b) What do you think were the likely effects on terrestrial and aquatic ecosystems along the lower part of the Colorado River after 1930?

2 Figure 9.19 shows the relationship between reservoir levels and earthquake frequency for (A) the Vaiont Dam, Italy; (B) the Koyna Dam, India; and (C) the Nurek Dam, Tajikistan. Answer the following questions.
  (a) Describe the correlation between reservoir construction and earthquake frequency.
  (b) Suggest reasons for the relationship described in part (a).
  (c) What are the environmental consequences of parts (a) and (b)?

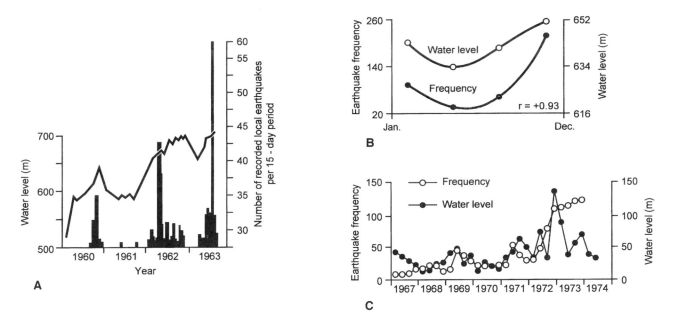

Figure 9.19 *The relationship between reservoir levels and earthquake frequencies for (A) the Vaiont Dam, Italy; (B) the Koyna Dam, India; (C) the Nurek Dam, Tajikistan. Redrawn after Goudie (1993b).*

Short questions

1 In the USA, what is the EPA?

2 What is laterisation?

3 Define the term desertification.

4 Describe the ecological effects of channelisation.

5 List the different types of human-induced threats to coral reefs.

6 What are environmental audits?

7 Describe the characteristics of thermokarst.

8 Outline the main human impact on the oceans and seas.

9 List the various ways that ground subsidence may occur.

10 Describe the various means to reduce soil erosion.

Answers to multiple-choice questions

1 c; 2 d; 3 b; 4 b; 5 a; 6 a; 7 a; 8 c; 9 d; 10 b; 11 b; 12 b; 13 b; 14 a; 15 b; 16 b; 17 d; 18 b; 19 c; 20 d; 21 c; 22 a; 23 c; 24 d; 25 d.

Answers to figure questions

1 (a) The changes in discharge and sediment load relate to dam construction. Sediment load decreases abruptly after 1930 because sediment carried in the Colorado River is allowed to settle out in the reservoir, whereas water is still allowed to flow downstream of the reservoir by controlled releases. (b) Reducing discharge and sediment loads in the Colorado River results in the loss of aquatic and nearshore habitats; as much of the river bed dries up, and riffles and pools that provide important niches are not as well developed or are destroyed.

2 (a) The frequency of earthquakes rises rapidly soon after water levels in reservoirs are increased. (b) The increased frequency of earthquakes associated with rising levels of water may be due to the increased stress exerted on the Earth's crust by the weight of the water. The higher ground-water levels allow the frictional stresses along the faults to be overcome because the water increases the pore fluid pressure, allowing the fault to rupture and cause an earthquake. (c) The environmental consequences range from damage associated with small earth tremors to the failure of the dam, which would result in catastrophic flooding.

Answers to short questions

1 The EPA is an abbreviation for the Environmental Protection Agency. It is responsible for managing federal efforts aimed at controlling air and water quality, reducing radiation and pesticide hazards, regulating the disposal of hazardous waste, and undertaking/sponsoring environmental research.

2 Laterisation is a tropical soil process that involves the enrichment of sesquioxides of aluminium and/or iron in a soil, which leads to the formation of lateritic layers, or a laterite soil. Lateritic layers are hard and inhibit plant growth, which in turn may lead to soil erosion and its associated problems.

3 Nelson (1988) provides one of the best definitions of desertification. He describes it as a process of sustained land (soil and vegetation) degradation in arid, semi-arid and dry sub-humid areas, caused at least partially by humans and reducing the productive potential to an extent that can neither be readily reversed by removing the cause nor easily reclaimed without substantial investment.

4 Channelisation of a stream can cause ecological degradation as it destroys vegetation and the stream drains more quickly, reducing the number of slack water habitats. This results in increased water temperatures since there is little vegetation to provide a cover. In addition, it results in rapid daily and seasonal fluctuations in temperature. This may affect many organisms that are sensitive to temperature changes. The higher stream velocities may be too great for some animals, reducing the number of habitats as the stream has mostly riffles and few pools. Furthermore, during periods of low flow the stream may dry up over much or all of its length, and would therefore not be able to support aquatic life.

5 Human-induced threats to coral reefs include: over-collecting of fish, giant clams, pearl oysters and coral; fishing using dynamite and poison; recreation, involving tourism, scuba diving and anchor damage; siltation due to erosion caused by fuelwood collection and deforestation; coastal development, involving causeway, road and housing construction, sand mining and dredging; pollution, involving oil and pesticide spillage, thermal and sewage; and military use, involving nuclear testing and conventional bombing.

6 Environmental audits provide a methodology for evaluating at regular intervals the environmental performance of companies, and for checking compliance with environmental regulations and codes of practice. Audits ensure that legislation is adhered to and that fines and litigation are thus prevented. They also help to create environmental awareness within the firm/company and improve its public image.

7 Thermokarst is the result of the degradation of permafrost. Subsiding waterlogged ground, and associated ponds and slurries, which are the result of melting ground ice, are characteristic.

8 The main human impact on the oceans and seas results from pollution by dumping and accidents, over-fishing, mineral extraction, and the removal of rare and important marine life.

9 Ground subsidence may be the result of extraction of solid materials such as coal, rock or salt; extraction of liquids such as the withdrawal of ground water or oil and gas; desiccation such as drying of peat bogs or sediments that contain swelling clays; and the formation of thermokarst.

10 Soil erosion control technologies reduce soil erosion. They include ridge planting; no-till cultivation; crop rotation; the use of grass strips; agroforestry; contour planting; the use of cover crops; and windbreaks.

Additional references

Board, P. 1996. Contaminated lands. *New Scientist*, 12 October, Inside Science, 94, 4 pp.

Daviss, B. 1996. Oracle of the oceans. *New Scientist*, 12 October, 26–31.

Holmes, B. 1996. The Amazon. *New Scientist*, 21 September (specially commissioned series of articles on many aspects of land use in the Amazon).

Aims

- To examine the main socio-economic and political issues surrounding environmental issues discussed in the textbook.

- To discuss strategies to help with the effective management of the global environment and to provide a manifesto for living.

- To examine the international conventions and legislation that is being undertaken to reduce the impact of environmental degradation.

Key point summary

- The four principal components of the ecosphere under threat are the climatic system, the nutrient cycles, the hydrological cycle and biodiversity.

- Population growth is a cause of major concern because of the stress that it imposes on the environment, although some argue that current world resources are capable of adequately sustaining an even larger global population.

- Enormous differences in wealth, life chances, health, education and social provision exist between the developed and less developed nations. In various parts of the world, such differences in access to the means of life have initiated wars and political instability, thereby contributing to environmental stress and degradation.

- The numbers of refugees are currently increasing and constitute not only a regional but also a major global problem.

- Urbanisation and population pressures have concentrated pollution, poor housing, disease and poverty into large megalopolises.

- In many cases, these social and environmental issues can be tackled only by international co-operation, the defrayment of much of the so-called 'Third World debt'.

- Agro-economic problems include over-intensive land use (for example, associated with the cultivation of industrial monocultures), inappropriate land use, the clearing of important natural vegetation, salinisation, laterisation, and pollution by fertilisers and pesticides such as nitrates.

- Methods to mitigate these effects of bad land management include farming practices that concentrate upon efficient but not over-intensive crop cultivation, less emphasis on the use of environmentally harmful fertilisers and pesticides, improving soil productivity, reducing soil erosion, and stopping salinisation and desertification.

- The rise of consumer society has led to an increased requirement for energy and natural resources. Without careful resource allocation and planning, there is a real danger that many resources may become severely depleted, something that could act as a limit to growth.

- As alternatives to conventional fossil fuels, renewable energy resources (for example solar, wind, wave, tide and biomass energy) should be encouraged, together with more research into technologies such as hydrogen energy.

- Without a concerted global effort to develop substantial energy supplies from renewables, there may be no alternative but to place greater reliance on nuclear energy, with its associated problems of radioactive waste disposal and the risk of major and long-term environmental pollution.

- The concept of 'sustainable development' was introduced in 1980 by the World Conservation Strategy, and the arguments developed by the International Union for the Conservation of Nature. In 1984, these international groups were absorbed into the World Commission on the Environment and Development, which produced the 1987 report *Our Common Future*, a document that ostensibly provided strategies for sustainable development. These strategies included reducing world poverty, improving agricultural practices, conserving energy, reducing anthropogenic greenhouse gas emissions, recycling waste, improving technologies, and reducing the disparities between rich and poor nations. The underlying arguments and strategies for sustainable development remain controversial.

- Atmospheric pollution has become a regional and global issue. International agreements and conventions on atmospheric pollution control resulted in the 1984 agreement to reduce sulphur emissions by 1993; the 1987 Montreal Protocol to reduce CFCs by 50 per cent by the year 2000, followed by a total ban on CFCs; the 1988 First World Conference on 'The Changing Atmosphere'; the 1989 Helsinki agreement on a total ban on CFCs by 86 countries by the year 2000; and the June 1992 United Nations 'Earth Summit' in Rio de Janeiro, where agreements and conventions were presented to preserve global biodiversity and to mitigate any possible global climate change precipitated by human activities. It was at Rio that Agenda 21 was signed by many nations. The Conference of Parties (COP), those who had ratified the United Nations Climate Change Convention at Rio de Janeiro in 1992, signed the Berlin mandate in April 1995 to return greenhouse emissions to 1990 levels by the year 2000, and to establish a working group directed by the IPCC to investigate strategies for reduced emissions after the year 2000.

- A manifesto for the management of the Earth is presented in order to stimulate debate. The manifesto represents our personal views on maximising the chances of achieving global sustainable development, reducing global pollution, eliminating poverty, and increasing the life chances of individuals wherever they are born. This manifesto includes: feeding the world and eliminating poverty; controlling population growth; improving basic medical care; expanding educational provisions at all levels; energy conservation; resource sharing; recycling materials and waste; international co-operation on global issues; reducing military expenditure; a non-nuclear future; efficient and environmentally sound farming practices; and preserving natural wildernesses.

Main learning hurdles

This chapter requires a knowledge of all the previous chapters. The instructor should, therefore, spend time reviewing the relative components of the appropriate chapters when discussing each of the topics. The students may be advised to review the book by reading the 'Key points summary' at the beginning of each chapter.

Key terms

Common Agricultural Policy (CAP); Comprehensive Environment Response, Compensation and Liability Act 1980 (CERCLA); fertility; General Agreement on Tariffs and Trade (GATT); infant mortality; life expectancy; Malthusian views; Organisation for Economic Co-operation and Development (OECD); polluter-pays principle; poverty; sustainable development; United Nations Environment Program (UNEP); World Commission on Environment and Development (WCED).

Issues for group discussion

Suggest what you think are the most effective methods in helping to reduce rapid population growth.
The discussion should focus on the problems of family planning and the reproductive freedoms of women. The discussion may then lead on to the debate over the carrying capacity of planet Earth and the view that population is a resource rather than a problem.

Discuss ways that the Earth's resources may be safely developed.
The discussion should examine the concept of sustainable development and the role of the UNEP and Agenda 21. The students should provide examples of successful management programmes.

Produce your own manifesto for the management of the Earth.
The students should produce their own strategies, in order of importance, and they should criticise the content of the manifesto in *An Introduction to Global Environmental Issues.*

Selected readings

Blum, E. 1993. Making biodiversity conservation profitable: a case study of the Merck/INBio agreement. *Environment*, 35 (4), 16–20 and 38–45.

This article provides an example of how business and the conservation of biodiversity can work together successfully.

Brundtland, G.H. 1994. The solution to a global crisis. *Environment*, 36 (10), 16–20.
This is a copy of Brundtland's speech to the World Population Conference in Cairo in 1994. It argues for the greater role of women in population control.

Chen, L.C., Fitzgerald, W.M. and Bates, L. 1995. Women, politics and global management. *Environment*, 37 (1), 4–9 and 31–3.
This article discusses the role of women in world development and management.

Cohen, J.E. 1995. Population growth and the Earth's human carrying capacity. *Science*, 21 July, 269, 341–6.
The paper examines if it is possible to calculate the number of people that can be supported on planet Earth.

Daily, G.C. 1995. Restoring value to the world's degraded lands. *Science*, 269, 350–4.
This paper discusses the remediation of degraded lands.

Frosch, R.A. 1995. The industrial ecology of the 21st Century. *Scientific American*, September, 178–81.
This article discusses the possible future trends in recycling of manufactured materials.

Glanz, J. 1995. EROS, MACHO, and OGLE net a haul of data. *Science*, 268, 642–3.
A useful review of the new satellites that will provide scientists with new ways of monitoring global environmental change.

Haas, P.M., Levy, M.A. and Parson, E.A. 1992. Appraising the Earth Summit: how should we judge UNCED's success? *Environment*, 34 (8), 6–12.
This is a comprehensive discussion of the successes and failures of the Earth Summit.

Hamer, M. 1993. City planners against global warming. *New Scientist*, 24 July, 12–13.
This is an article that considers the design of cities in the light of global warming.

Livernash, R. 1995. The future of populous economies: China and India shape their destinies. *Environment*, 37 (6), 7–11 and 25–34.
This article discusses the characteristics and problems associated with the population growth of the two most populous countries in the world.

Nowak, M.A. and McMichael, A.J. 1995 How HIV defeats the immune system. *Scientific American*, August, 58–65.
This is a comprehensive summary of the mechanics of HIV and the immune system, advocating a theory of continuous mutation of the HIV virus.

Pearce, F. 1994. Soldiers lay waste to Africa's oldest park. *New Scientist*, 3 December, 4.
This article illustrates the effects war, in this case the Rwandan civil war, and refugees can have on environmental degradation.

Plucknett, D.L. and Winkelmann, D.L. 1995. Technology for sustainable agriculture. *Scientific American*, September, 182–6.
An interesting article proposing that agricultural technologies will develop over the next century that will focus on sustainable development.

Root, T.L. and Scheider, S.H. 1995. Ecology and climate: research strategies and implications. *Science*, 21 July, 269, 334–40.
This paper submits possible research strategies that will be useful in examining climate and ecological change.

Thompson, D. 1995. Problems of plenty. *Time*, 21 August, 22–3.
This is an interesting journalist article illustrating the problems of poor infrastructure and management in the production of food in India.

Textbooks

Adams, W.M. 1990. *Green Development: Environment and Sustainability in the Third World*. London: Routledge.
This is a book on the problems of development and its environmental impact in the developing ['Third'] world. This book addresses the problems of striving for sustainable development and provides an important read for students and instructors of development and environmental studies.

Brandt, W. 1980. *North–South: A Programme for Survival*. The report of the Independent Commission on International Development Issues under the Chairmanship of Willy Brandt. London: Pan Books.
Brandt, W. 1983. *Common Crisis North–South: Co-operation for World Recovery*. The Brandt Commission 1983. London: Pan Books.
These books are important historical documents. They are the result of investigations by a group of

international leaders into the problems of inequality in the world and the failure of economic systems to tackle the issues. The group proposes a spectrum of bold recommendations and reforms to avoid the perceived imminent world economic crisis. The authors describe different elements of the global crisis in trade, energy and food supply, concentrating on the overriding problem of how to provide the finance for help, and on ways of compensating for the decline in financial liquidity to reverse the decline in trade and to raise the overall world economy. The WCED followed these reports with its publication of *Our Common Future* (see below). These books are essential for all students, instructors and policy-makers with concerns in global environmental issues.

Ekins, P. 1992. *A New World Order: Grassroots Movements for Global Change*. London: Routledge.
A thought-provoking book on the problems associated with and resulting from war, insecurity and militarisation, poverty, the denial of human rights, and environmental destruction. Attention focuses on the possible solutions to these problems at a grassroots level.

Johnson, R.J., Taylor, P.J. and Watts, M. 1995. *Geographies of Global Change*. Oxford: Blackwell.
A useful book that explores geo-economic, geo-political, geo-social, geo-cultural and geo-environmental change. The book considers the collapse of socialism; the reconfiguration of North Atlantic capitalism; the hypermobility of capital; the rise of ferocious nationalisms; global environmental change; the power of international media; the social movements associated with population growth; and international migration. It provides a useful economic, political, social, cultural and ecological view of change at every geographical scale from the global to the local. It is an essential text for undergraduate and graduate students of geography.

May, P. (ed.) 1996. *Environmental Management and Governance*. London: Routledge.
This book examines aspects and problems of environmental management. It considers the role of governments, both at local and national levels, and the strengths and weaknesses of co-operative versus coercive environmental management, through a focus on the management of natural hazards. It presents new and innovative environmental management and planning programmes, with particular focus on North America and Australia.

Mikesell, R.F. 1995. *Economic Development and the Environment*. London: Cassell.

This book examines how the environment and sustainability integrate with development programmes and strategies. It outlines the conceptual and theoretical issues involved in sustainable development, as well as providing case studies that compare the successfulness of various types of development project.

Moore Lappe, F. and Schurman, R. 1989. *Taking Population Seriously*. London: Earthscan.
This book provides a useful analysis of the reasons for population growth. The authors discuss the need to understand the underlying social and economic causes of population growth in order to implement effective population control.

Nebel, B.J. and Wright, R.T. 1993. *Environmental Science: The Way the World Works*, fourth edition. Englewood Cliffs, NJ: Prentice Hall.
A well-written and illustrated textbook, containing review questions and other exercise sections at the end of each chapter. This book is divided into four parts, dealing with (1) what ecosystems are and how they work; (2) finding a balance between population, soil, water and agriculture; (3) pollution; and (4) resources: biota, refuse, energy, and land. There is a useful bibliography and glossary at the back of this book. It is a useful text for college students and instructors in environmental sciences.

Omara-Ojungu, P.H. 1992. *Resource Management in Developing Countries*. Harlow: Longman.
This text examines the problems of resource management in developing countries, outlining the basic ecological, economic, technological and ethnological aspects of resource management. It emphasises that poverty is the critical problem facing resource management and development. Examples are provided from Africa, Southeast Asia and Latin America.

Redclift, M. 1987. *Sustainable Development: Exploring the Contradictions*. London: Methuen.
This book argues that the development recommendations of the WCED report (WCED 1987) need to be redirected to give greater emphasis to local (indigenous) knowledge and experience if effective political action is to be taken to minimise any environmental damage. A book that is easily read, containing many interesting examples and recommendations.

Sarre, P. (ed.) 1991. *Environment, Population and Development*. London: Hodder & Stoughton, 304 pp.

A British Open University text that examines environmental issues with reference to population growth and economic and technological development. It is well illustrated with a good introduction for students concerned with environmental issues. Topics dealt with include population dynamics, agriculture, productivity and sustainability, urbanisation, behaviour, and social problems.

United Nations 1994. *World Programme of Action: International Conference on Population and Development*. New York: United Nations Population Fund. This is an important document resulting from the International Conference on Population and Development in Cairo in 1994. The book comprises 16 chapters. The first two chapters contain the preamble and principles, setting the overall tone of the document. Chapter 3 discusses general development while Chapters 4 to 8 deal with the core issues involving the empowerment of women, families, under-served groups, and reproductive and sexual health and rights. Chapters 9 and 10 discuss migration, while Chapters 11 and 12 deal with education, and technology, research and development, respectively. Chapters 13 through to 16 address national and international action, finances, and relationships with nongovernmental organisations. It is essential reading for all concerned with development issues.

World Commission on Environment and Development, 1987. *Our Common Future*. Oxford: Oxford University Press.
The report of the World Commission on Environment and Development examines the critical environmental and developmental problems, and contains many very useful tables and figures. Emphasis is placed on the economic and ecological factors that may lead to sustainable development. This is an essential reference source for all college and university students, instructors and policy-makers concerned with environmental and development issues.

Essay questions

1 Is sustainable global economic growth feasible, and even desirable?

2 Consider the assertion that the single biggest problem facing the survival of humankind is over-population.

3 'There is a North–South divide in the world between the rich and industrialised nations and the poorer, less industrialised nations. The only way to rectify this inequality of wealth and life chances is to make the rich nations so well off and comfortable that they will freely provide the aid necessary to raise the global standard of living.' Discuss the arguments for and against these statements.

4 Describe the possible methods of helping to alleviate poverty in developing countries.

5 Does society seriously undervalue education as a means of eliminating poverty and reducing disease?

6 'The majority of world's population is poor because there are too many people and not enough resources.' Discuss.

7 Describe the environmental consequences of war.

8 To what extent should international policy on environmental issues take cognizance of various cultural and religious groups?

9 Describe the characteristics of urban pollution and assess the various ways in which the effects may be mitigated.

10 'Women's education is the single most important route to higher productivity, lower infant mortality and lower fertility.' Discuss.

11 'Environmental awareness is an empty posture if it is not supported by an understanding of the economic, social and political context of environmental issues.' Discuss.

12 'Management of the environment is largely concerned with social, economic and political issues'. Evaluate this statement.

13 Examine the argument that the environment is the unnoticed victim of the debt crisis in the Third World.

14 Evaluate the contention that the principles and practices of environmental sustainability are insufficient to meet the challenges posed by economic growth.

15 Examine the key variables controlling population growth and decline, illustrating your answer with examples taken from both the developed and Third Worlds.

16 'While the goals and principles of sustainable development have been defined, the move from principle to practice is far from easy, and poses a series of dilemmas for which there are no clear solutions.' Discuss.

17 'To talk of global politics is to acknowledge that political activity is no longer primarily defined by national, legal and territorial boundaries.' Discuss the implications of this statement with reference to the environment.

Multiple-choice questions

Choose the best answer for each of the following questions.

1 The population of the world today is more than:
    (a) 1 billion
    (b) 5 billion
    (c) 10 billion
    (d) 15 billion

2 The world population has doubled since:
    (a) the beginning of this century
    (b) the late 1950s
    (c) the 1970s
    (d) the beginning of this decade

3 *The Ultimate Resource* was written by:
    (a) Reverend Thomas Malthus
    (b) Moore Lappe and Schurman
    (c) William C. Clark
    (d) Julian Simon

4 The view that accelerated population growth will exceed a critical level, such that the scarcity of food would cause famine and war, thereby reducing the population to some sort of equilibrium level, is known as a:
    (a) Darwinian view
    (b) Marxist view
    (c) Malthusian view
    (d) creationist's view

5 AIDS was first recognised in:
  (a) 1981
  (b) 1982
  (c) 1983
  (d) 1984

6 Mexico City, which is the largest city in the world, has a population of more than:
  (a) 7 million
  (b) 19 million
  (c) 23 million
  (d) 27 million

7 The UN-sponsored International Conference on Population and Development (ICPD) in Cairo in 1994 produced a 113-page document outlining a new population policy that aimed to stabilise global population by 2015 at about 7.27 billion. This is known as:
  (a) the World Programme of Action
  (b) *Our Common Future*
  (c) Agenda 21
  (d) the Sustainable Development of Population

8 The UN-sponsored International Conference on Population and Development (ICPD) in Cairo in 1994 emphasised that one of the most important processes to strive towards a sustainable population is:
  (a) increased women's rights
  (b) increased resource exploitation
  (c) decreased energy use
  (d) decreased resource exploitation

9 'Selective primary health care' is a broad strategy that focuses on:
  (a) prevention, curative and rehabilitative services to whole communities
  (b) the monitoring of the growth of children, the use of oral rehydration salts for diarrhoea, the encouragement of breast feeding for infants, and immunisation
  (c) the eradication of contagious diseases by improved research and education
  (d) the immunisation of populations in developing countries

10 The Montreal Protocol, in 1987, called for a:
  (a) 30 per cent reduction in CFCs by the year 2000
  (b) 90 per cent reduction in CFCs by the year 2000
  (c) total ban on CFCs by the year 2000
  (d) total ban on CFCs by the year 2025

11 The Conference of Parties (COP), those who had ratified the United Nations Climate Change Convention at Rio de Janeiro in 1992, ratified the Berlin mandate in 1995, which:
  (a) agreed a total ban on greenhouse emissions by the year 2000
  (b) agreed a reduction of greenhouse emissions to 1990 levels by the year 2000
  (c) agreed a reduction of greenhouse emissions to 1980 levels by the year 2000
  (d) agreed a reduction of greenhouse emissions to pre-Industrial Revolution levels by the year 2000

12 IPCC is the abbreviation for:
  (a) International Panel on Climate Change
  (b) Intergovernmental Panel on Climate Change
  (c) Independent Panel on Climate Change
  (d) International Panel on Climate Crisis

13 The International Conference on an Agenda for Environment and Development into the 21st Century (ASCEND 21) was convened in:
  (a) Vienna during November 1991
  (b) Rio de Janeiro during June 1992
  (c) Berlin during April 1994
  (d) Beijing during September 1995

14 On 3 June 1992, more than 100 world leaders and 30,000 other participants met at an extraordinary meeting in Rio de Janeiro for the beginning of the Earth Summit. This was also known as the:
  (a) United Nations Conference on Environment and Development
  (b) ASCEND 21 conference
  (c) Rio Conference
  (d) International Conference on Climate Change

15 The concept of 'sustainable development' was introduced in 1980 in:
  (a) the World Conservation Strategy
  (b) the WECD report, *Our Common Future*
  (c) Agenda 21
  (d) ASCEND 21

16 The maximum population that can be maintained by a habitat or ecosystem without degrading the ability of that habitat or ecosystem to maintain that population in the future is known as the:
  (a) carrying capacity
  (b) sustainable capacity
  (c) limits to growth
  (d) a, b and c

17 Who wrote *Limits to Growth*?
  (a) Malthus
  (b) Meadows *et al.*
  (c) Houghton *et al.*
  (d) Simon

18 In 1981, J. Simon wrote:
  (a) *The Ultimate Resource*
  (b) *The Population Bomb*
  (c) *Limits to Growth*
  (d) *Taking Population Seriously*

19 The World Wildlife Fund estimates that humans are still exterminating vertebrates at a rate of:
  (a) 1 per cent of world species annually
  (b) 2 per cent of world species annually
  (c) 5 per cent of world species annually
  (d) 10 per cent of world species annually

20 The richest 15 per cent of the world's population consumes:
  (a) 10 per cent of the world's energy
  (b) 20 per cent of the world's energy
  (c) 50 per cent of the world's energy
  (d) 90 per cent of the world's energy

21 The number of refugees in Africa today is approximately:
  (a) 1 million
  (b) 2 million
  (c) 4 million
  (d) >5 million

22 After the Soviet withdrawal from Afghanistan in 1988 there were approximately:
  (a) 100,000 land mines left behind
  (b) 1 million land mines left behind
  (c) 5 million land mines left behind
  (d) 10 million land mines left behind

23 The major source of urban air pollution is:
  (a) industrial processes
  (b) coal-fired power stations
  (c) vehicle emissions
  (d) leaking gas pipes

24 Which of the following countries does not attend G7 summits?
  (a) USA
  (b) Denmark
  (c) Germany
  (d) France

25 Which of the following statements is false?
  (a) the female literacy rate in developing countries is still only two-thirds that of males
  (b) a sixth of the people in developing countries still go hungry every year
  (c) the average life expectancy in developing countries is still 12 years shorter than in developed countries
  (d) the maternal mortality rate in developing countries is twice that of developed countries

Figure questions

1 Figure 10.3 shows the age distribution of populations in 1990 compared with that for 2025. Answer the following questions.
  (a) Explain the difference between the distributions in less developed and developed countries for 1990.
  (b) List ways in which the rate of population growth may be reduced.

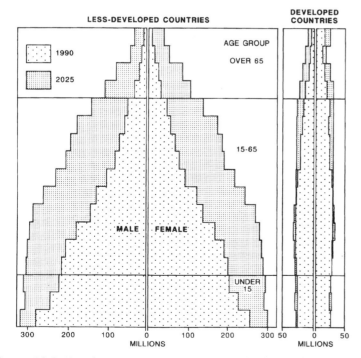

Figure 10.3 *Age distribution of populations of the less developed and the developed countries in 1990 compared with that projected for 2025. In the less developed countries, the population will continue to grow rapidly, with an expanding labour force. The percentage of old people will also increase with respect to the young, requiring greater care. Redrawn after Keyfitz (1989).*

2 Figure 10.6 shows energy intensity versus time in industrialised and developing countries. Answer the following questions.

(a) Describe the characteristics of the energy intensity throughout time for developed countries and compare these characteristics with those projected for developing countries.

(b) List the likely problems that are associated with the growth of energy intensity in developing countries.

(c) Suggest possible means to reduce the problems highlighted in (b).

3 Figure 10.17 shows the growth in surface transport: movement of people by mode 1952–93 in the UK for car/van; bus/coach; rail; cycles. Answer the following questions.

(a) What is the likely environmental consequence of the rapid growth shown by the solid continuous line?

(b) Is this trend likely to continue? Explain your answer.

(c) List the various ways in which this trend might be reduced.

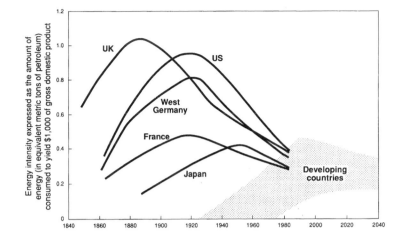

Figure 10.6 *Energy intensity versus time in industrialised and developing countries.*

Answers to multiple-choice questions

1 c; 2 b; 3 d; 4 c; 5 a; 6 b; 7 a; 8 a; 9 b; 10 c; 11 b; 12 b; 13 a; 14 a; 15 a; 16 a; 17 b; 18 a; 19 a; 20 c; 21 d; 22 d; 23 c; 24 b; 25 d (it is 122 times).

Short questions

1 Outline the concept of sustainable development.

2 What is the 'Superfund'?

3 Compare the population structure of developed and developing countries.

4 What is GATT?

5 Describe the types of environmental degradation that refugees cause.

6 What is a Malthusian view?

7 Summarise the main finding of the *Limits to Growth* report, which was presented to the Club of Rome in 1972.

8 List the main recommendations proposed by Pickering and Owen in their 'Manifesto for Living'.

9 What was the Earth Summit and what is Agenda 21?

10 Describe the main problems associated with agro-economics.

Answers to figure questions

1 (a) In developing countries, the population is dominated by the younger age groups, while in developed countries the population within each age group is very similar. This pattern is due to the high mortality rates in the younger age groups in less developed countries as compared with developed countries. (b) Reducing population is a difficult problem but the following methods may help: family planning programmes; provisions for women's education; and increased wealth.

2 (a) The graphs show that as countries become industrialised the energy intensity increases, after which it drops rapidly. Developing countries are undergoing similar trends but the decrease is not as dramatic as for the G7 countries shown on the diagram. (b) The main problems associated with such growth are increased pollution and associated environmental health problems, as well as a lowering of the quality of life for ordinary people living in industrial regions. (c) The best way to reduce the problems highlighted in (b) is to reduce the amounts of emissions by improved clean-energy technologies.

3 (a) The likely environmental consequences of this growth in surface transport is increased pollution both

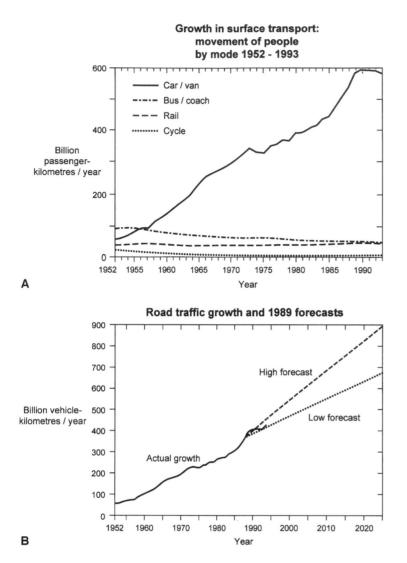

Figure 10.17 *(A) Growth in surface transport: movement of people by mode 1952–93 in Great Britain; and (B) road traffic growth and 1989 forecasts for Great Britain. Redrawn after Royal Commission on Environmental Pollution (1994).*

locally and regionally, as well as globally from exhaust emissions. (b) The trend is unlikely to continue, because the graph plateaus at about 1988; however, this may be a temporary check in the growth. (c) The trend is likely to decrease only if governments provide adequate public transportation systems and incentives to use them rather than private vehicles.

Answers to short questions

1 Sustainable development is a concept that strives for the development of available resources without compromising the ability of future generations to meet their needs.

2 The Superfund is the popular name for the US Comprehensive Environmental Response, Compensation, and Liability Act (CERCLA) of 1980, which provides a mechanism and funding to help to clean up potentially dangerous hazardous waste sites.

3 In developing countries, the population is dominated by the younger age groups, while in developed countries the population within each age group is very similar.

4 GATT is an acronym for 'General Agreement on Tariffs and Trade', which is an international mechanism to control and regulate economic growth through legislative/fiscal policy.

5 Any major diaspora may lead to environmental degradation. This is due to a number of factors, e.g. the sudden increase in environment stress associated with the need for food and fuel in a new region and/or neighbouring country, possibly resulting in vegetational destruction and soil erosion. Also, large numbers of people may pollute water resources due to poor sanitation.

6 A Malthusian view is that population increases faster than resources, with the result that population growth is checked by famine, plagues and other problems. Neo-Malthusians also include environmental degradation as a consequence of over-population. These views follow the beliefs of Reverend Thomas Malthus, who proposed such an idea in 1789 in his *An Essay on the Principle of Population and a Summary View of the Principle of Population.*

7 The *Limits to Growth* report claims to show that the existing patterns of global resource use will lead to a collapse of the world's socio-economic and political systems within the next century.

8 The main recommendations proposed by Pickering and Owen in their 'Manifesto for Living' include: feeding the world and eliminating national poverty; control of population growth; expanding educational provision at all levels; energy conservation; resource sharing; recycling resources and materials; international co-operation on global issues; reducing military expenditure; working towards a non-nuclear future; ethical investments; practising efficient and environmentally sound farming; and leaving designated natural wildernesses undeveloped and unexploited.

9 The Earth Summit is the popular name for the United Nations Conference on the Environment and Development, which was held in Rio de Janeiro, Brazil, 3–14 June 1992. Agenda 21, the Rio Declaration on Environment and Development, was presented at the Earth Summit. It aims to providing a blueprint for global action into the twenty-first century that could be agreed by the attending delegates.

10 Agro-economic problems include over-intensive land use (e.g. associated with the cultivation of industrial monocultures); inappropriate land use; the clearing of important natural vegetation; salinisation, laterisation; and pollution by fertilisers and pesticides such as nitrates.

This section describes alternative methods of assessment and ideas for assignments. It is best to test the student's ability in as many ways as possible, remembering that not all students can write essays well or give good presentations. Most environmental scientists will have to undertake a variety of tasks, including fieldwork, report writing, literature reviews, data processes and giving presentations, so it is therefore important to develop all these skills.

Poster presentation

A poster presentation is an interesting and exciting way to assess students. A poster tests the student's ability to research topics, and to assimilate information and present it in a clear and concise manner. Our experience has shown that students value and enjoy this type of exercise and generally perform better than in equivalent essay assignments. The ability to be able to present evidence and argue issues clearly and effectively is an important skill that students must develop.

Students should be given an appropriate topic with instructions on the poster format and the presentation (see next section). The instructor should emphasise the importance of research, referencing on the posters, clarity, presentation and content. The talks should not be more than about 15 minutes. This encourages the student to be concise and helps the student to focus on the main points. They should prepare clear overheads and/or slides and they should practise their talks before they are assessed. There should also be an opportunity for the students and instructors to ask questions. It is useful for the students to see examples of display boards, and instructors are encouraged to keep a collection of posters from previous classes. The students should be told how they are going to be assessed. This will help them to appreciate how to construct a poster presentation. A possible marking scheme is shown on the assessment form (see page 105). This will also help the instructor to assess the talks and posters as objectively as possible. Instructors may use the following useful outlines.

Poster presentation instructions

Prepare a 15-minute talk and a display board on one of the topics listed below. Your display board should not be bigger than 90 cm × 60 cm. Make your board and presentation clear, summarising the main aspects of the topic. Make full use of the available space, but do not overcrowd it with a small typeface; 12-point font is a convenient size. You should focus on summarising the important points, illustrating them with figures and tables. Reference sources must be listed on the boards, on the diagrams and in the text.

Your presentation should be clear and audible. It should summarise the main points that you may have included on your display board. Use overheads and slides to illustrate your talk. Do not write smaller than 18 point on overheads and do not use more than about 20 words per overhead. Make full use of colour graphs and summary diagrams. Avoid using tables if you can present the data as graphs. Practise your talk to help you with timing; it will also help to give you confidence. Any person talking for more than 15 minutes will be stopped by the chairperson of each session.

For further discussion of the use of poster exercises the instructor should read:

Howenstine, E., Hay, I., Delaney, E., Bell, J., Norris, F., Wheln, A., Pirani, M., Chow, T. and Ross, A. 1988. Using a poster exercise in an introductory geography course. *Journal of Geography in Higher Education*, 12 (2), 139-47.

ASSESSMENT FORM

Name: _____

Course/Year: _____

Title: _____

_____

_____

Date: _____

Presentation

Content, coverage and understanding of topic (25%):   _____

Research and originality (15%):   _____

Presentation (overheads, slides, etc.) (10%):   _____

Subtotal (50%)   _____

Poster

Content, coverage and understanding of topic (25%):   _____

Research and originality (15%):   _____

Presentation (overheads, slides, etc.) (10%):   _____

Subtotal (50%)   _____

Total (100%)   _____

Assessors – First marker: _____

Second marker: _____

Agreed overall mark: _____

Environmental impact assessments

An environmental impact assessment (EIA) is a useful way for students to develop a variety of technical skills as well as to develop an ability to work within a group. An EIA helps to evaluate the environmental acceptability of projects under consideration. The technique involves an amalgamation of studies based on predetermined approaches. For an EIA to succeed there needs to be an integration of the disciplines. To be most effective, students should arrange themselves into groups of between 3 to 5. The composition of the group depends heavily on the nature of the class being instructed. For example, the group might comprise geologists, biologists, chemists, etc. The instructor should encourage the groups to have a mix of disciplines. The subject of the EIA should be clearly outlined to the students and the students should be allowed to assign themselves to particular tasks dependent upon their particular speciality. Each student should write individual reports, and report back to the group where they collate their work as a summary report. Possible examples of subjects for EIAs include:

1 A highway construction
2 The construction of a marine barrage across an estuary
3 The construction of a landfill site
4 The construction of a utility such as a hospital, power station or amusement park
5 The construction of a bridge
6 The reclamation of wetland for agriculture

The instructor should identify a real geographical location for one or more of the above projects. The students will then be able to obtain the relevant information on, for example, the subjects listed below:

1 Geological data
2 Ground-water and surface-water data showing quality and routes of flow
3 Soil data showing quality and composition
4 Geomorphological data, including topographic, flood, landslide and erosion data
5 Biological data, showing important or unique floras or faunas and the role of vegetation in other aspects of the environment within the region
6 Climatic data
7 Locations of protected sites such as Sites of Special Scientific Interest or Heritage Sites
8 Population distributions

Individual reports should present these aspects after the students have chosen the relevant topic. *An*

Introduction to Global Environmental Issues discusses each of these topics with relevant references, and the instructor should encourage the student to read the appropriate text and references. Once all the various aspects have been considered, the students should write a summary report for the EIA. The instructor is encouraged to get the students to present their results orally. The students should be given examples of previous EIAs, if possible, and the method of assessment should be outlined. Marks should be given for the specialist report undertaken by individual students, the group summary and the oral presentation.

Fieldwork

Fieldwork is an essential component of the environmental sciences. It is a natural laboratory for the collection of data, in which to acquire information that constrains theoretical models and to test hypotheses. Environmental impact assessments are developed from data gathered in natural environments.

The nature of appropriate fieldwork will depend largely on the speciality of the group of students. Fieldwork includes geological mapping, slope stability analysis, soil surveys, vegetation inventories, sampling waters, soils and rocks for geochemical work, hydrogeological surveys, monitoring weather conditions, air-quality tests, traffic monitoring, and a panoply of social surveys. It is beyond the scope of this manual to describe all the types of fieldwork that can be undertaken. We emphasise, however, the importance of field instruction as an essential means of teaching students skills, providing direct experience of the natural world, and perhaps most importantly as the only laboratory in which to test models and define sensible boundary conditions that go into theoretical models. Fieldwork teaches everyone, student and teacher alike, that the natural world involves many complex variables, with widely differing degrees of uncertainty attached to any observations and inferences, and that theoretical models are no more than idealised abstractions and simplifications of natural phenomena.

In any fieldwork programme, the lecturer should encourage students to record systematically appropriate data in statistically significant data sets. As with any analytical study, students should define the magnitude of any errors, both machine and instrument based, and attempt a quantitative estimate of other errors, e.g. human error in recording data, etc. Where appropriate, field-based observations should be subjected to statistical analysis and, therefore, any course in the environmental sciences should include some formal training in statistics.